Transactions
of the
American Philosophical Society
Held at Philadelphia
For Promoting Useful Knowledge
Volume 91, Part 2

"Fighting for the Good Cause"
Reflections on Francis Galton's Legacy to American Hereditarian Psychology

Gerald Sweeney

American Philosophical Society
Independence Square • Philadelphia
2001

ISBN: 0-87169-912-5
US ISSN: 0065-9746

Library of Congress Cataloging-in-Publication Data

Sweeney, Gerald, 1942-
Fighting for the good cause : reflections on Francis Galton's legacy to American hereditarian psychology / Gerald Sweeney.
p.   cm.-- (Transactions of the American Philosophical Society; v. 91, Pt. 2) Includes bibliographical references and index.
ISBN 0-87169-912-5 (pbk.)
1. Genetic psychology. 2. Nature and nurture. 3. Heredity, Human. 4. Galton, Francis, Sir, 1822-1911. I. Title. II. Series.

BF70l .S93 2001
155 .2'34'0973--dc21   2001022674

CONTENTS

List of Illustrations

For Susy

ABSTRACT

"Fighting for the Good Cause": Reflections on Francis
Galton's Legacy to American Hereditarian Psychology

Sir Francis Galton is well understood to have served as an influential mentor for the educational psychologists who supplied crucial doctrine to American eugenics in its classic period, 1903 to 1930. Yet the nature of his influence has never been specified. The psychologists' own claim as to Galton's contribution—that he provided sufficient justification for their absolutist hereditarianism—was clearly disingenuous. Rather, the English polymath appears to have functioned in large part as a model for these figures, who appear to have been instrumentally informed by their perceptions of Galton's ulterior purposes in constructing eugenics as he did. Any of various features in the forty-five-year-long course of that development could have encouraged these particular legatees to appreciate both Galton and his product as surreptitious stanchers of democracy.

These psychologists, who were forced to seek after Galton in his writings, seem to have had good access to his wide-ranging promotions of eugenics, along with an understanding of the deficiencies contained in his and Karl Pearson's argument. Accordingly, Galton looks not so much to have transformed them into the promoters of eugenics they became as to have shown them why and how to become such, when other developments led them in that direction.

ACKNOWLEDGMENTS

I greatly appreciate this opportunity to acknowledge those who have helped with the work on this project and the larger undertaking of which it forms a part. Good friends have helped as only they can, by supplying the world worth returning to that solitary pursuits depend so much upon. These dear friends have included, through the years, the late Javad Emami, Bob Yale, Larry Shaw, Bryan Mack, John Dennis, Jonathan Petropoulos, David Leviatin, Brett Flehinger, Michael Pak, and Cliff Rutt. The list of those who have rendered selfless professional assistance begins with my professors at San Diego State University and includes William Cheek, Thomas Cox, Charles Hamilton, Raymond G. Starr, Kingsley Widmer, (an exchange professor) Willem van Hoorn, and especially the late Karl Keller, Jerome Sattler, Richard Steele, and Howard I. Kushner. These last two mentioned, I should note, have accorded various of my productions the most exhaustive and insightful scrutiny they have received.

At Harvard I had the honor of eliciting the encouragement and counsel, and in many cases the critical attention of, in the Department of History, Stephan Thernstrom, Akira Iriye (who has provided his dependably kind support on several levels), Alan Brinkley, and my dear advisor, mentor, and friend Donald Fleming. Professor Fleming's scholarly guidance and example have been coupled with his legendary generosity, patient forbearance, and unwavering support. In the Department of Psychology, I have received good advice and kind encouragement from Brendan Maher and his wife and academic partner Barbara. With respect to the Department of the History of Science, I have profited greatly from the encouragement and critiques of my work provided by Everett Mendelsohn, Loren Graham, Barbara Rosenkrantz, and visiting scholars Dorothy Porter and Pietro Corsi. Other scholars in the field of the history of psychology who have been especially generous with their time and support include Raymond E. Fancher, James Reed, William R. Woodward, Michael M. Sokal, and Richard T. von Mayrhauser. To return to Harvard's History Department more generally, those who rendered vital assistance at critical points have included Cory Paulsen, Geri Malatesta, and the past departmental chairs Angeliki Laiou, who provided both encouragement and quite helpful summer research grants, and Thomas Bisson and William Kirby, who supplied money for research travel to England much later in my graduate student tenure than is normally done. I am also pleased to acknowledge the ever-timely support and interventions on my behalf of Sarah White, then a Financial Aid officer, who paid me the supreme compliment of reading my Master's thesis in its entirety, and discussing it with me at length. I also express my considerable gratitude to the Charles Warren Center for Studies in American History, which, under the direction of Bernard Bailyn, extended several quite generous research grants, and has subsequently, under the direction of Laurel Thatcher Ulrich, afforded me a number of nonresident fellowships, which have helped most instrumentally with the furtherance of my work overall. Particularly I would like to thank Susan Hunt, the Center's longtime Administrator and institutional anchor, for her many kindnesses and helpful advice.

It is a particular pleasure to express gratitude to my editor at the American Philosophical Society, Carole Le Faivre-Rochester, whose advice and encouragement have meant much, and whose editorial assistance has been both heroic and edifying, and to Susan Babbitt.

Certainly, I wish to thank my family—my beautiful daughter Blake Kahan, my loving siblings Judith Carroll, Jim Sweeney, and John Sweeney, my dear father James Sweeney, who always believed in me, and my recently-departed mother Faith Sweeney, whose unflagging solicitude and tender comradeship bolster me still. Lastly, I can only attempt to acknowledge and thank the one to whom this small effort is dedicated, my dear wife Suejung Ko, who has supported me in all ways imaginable, always with the same fierce gladness and serene equanimity that I have treasured in her from our very beginning together.

Finally, I find it unusually requisite to engage in the time-honored tradition of absolving all who have helped of any responsibility for what I have written here. Almost all who have perused this statement in its several iterations have expressed emphatic reservations as to various of its features, and only a few of these perceived pitfalls have been filled fully in. My only defense for such obduracy is that I have hoped that this treatment of Galton—which is greatly more reliant on inference than the work it is intended to preface—might touch off a more searching consideration of the man, his invention, and its purposes than they thus far have received.

CHAPTER ONE
INTRODUCTION

SIR FRANCIS GALTON towered over the pantheon of authorities cited by the educational psychologists at the doctrinal controls of American eugenics from 1903 to 1930. The greatest debt these hereditarian psychologists acknowledged to the English polymath was for having proven, ostensibly, their crucial common claim that heredity hugely predominates over environment in the making of individual intelligence. Edward L. Thorndike, the ranking educational psychologist of the period, assured readers of his first educational psychology text, published in 1903 and greatly expanded in 1914, that Galton's rigidly hereditarian ruling on the nature-nurture question, "that of an eminently fair scientific man based on an extensive study of individual biographies, may safely be taken with a very slight discount." Lewis M. Terman, the foremost developer of intelligence tests, indicated the basal quality of Galton's support in 1922: "All the available facts that science has to offer support the Galtonian theory that mental abilities are chiefly a matter of original endowment," he declared.[1]

Certainly these figures owed Galton a conspicuous doctrinal debt. The intrepid explorer of Africa, meteorology, anthropology, psychology, heredity and statistics had invented eugenics entirely singlehandedly, in 1865. He had issued its basic claims, coined its vocabulary, conferred its rules, and devised most of the statistical and experimental techniques it would require. He had even brought forth the new conception of heredity that orthodox eugenics depended upon—the modern counter to Lamarckism more success-

---

[1] Thorndike, *Educational Psychology* (New York: Science Press, 1903), 53; Thorndike, *Educational Psychology*, vol. 3, Mental Work and Fatigue, and Individual Differences and Their Causes (New York: Teachers College, Columbia University, 1914), 236; Terman, "Were We Born That Way? Or Can We Help It? Is Heredity or Environment the Power that Moulds Us? What Science Now Knows About Intellectual Differences, And Their Significance," *World's Work*, October 1922; 657 (misnumbered 659). See also Terman, "The Psychological Determinist; or Democracy and the I. Q.," *Journal of Educational Research* 6 (June 1922): 60. For a representative eugenics propagandist attesting to the power of Galton's evidence, see Lothrop Stoddard, *The Revolt Against Civilization: The Menace of the Under Man* (New York: Scribner and Sons, 1922), 48-50.

fully advanced by August Weismann nearly twenty years later.

Moreover, indications abound that Galton served as some form of model for these psychologists. "The writings of James and Galton have influenced me most," Thorndike explained in an auto-biographical sketch. Terman divulged in his autobiographical entry that "of all the founders of modern psychology, my greatest admiration is for Galton." The English emigré and preeminent instinct psychologist William McDougall demonstrated the depth of Galton's impress on his thinking in his "A Practicable Eugenic Suggestion," given before the (English) Sociological Society in 1906.[2]

Yet if it has been widely observed that Galton exercised a formative influence on American hereditarian psychology, it remains to specify the nature of the doctrine, model, and proofs provided. Certainly, attempting to grasp the character of the contri-butions supplied by Galton seems a useful preliminary to comprehending what these hereditarian psychologists did in their classic period of influence. Arguably, it is difficult to understand what they were doing—and hoping to do—without first acquiring some sense of their view of what their English mentor had been about before them. All the same, the first issue to be addressed is the extent of Galton's success in supplying proof to his American heirs for their basic shared claim that heredity is nearly all and environment almost nothing in the making of individual intelligence.

---

[2] Thorndike, "Edward Lee Thorndike," in *A History of Psychology in Autobiography*, vol. 3, ed. Carl Murchison (Worcester, MA: Clark University Press, 1936), 268; Terman, "Lewis M. Terman [,] Trails to Psychology," in *A History of Psychology in Autobiography*, vol. 2, ed. Carl Murchison (Worcester, MA: Clark University Press, 1932), 331; McDougall, "A Practicable Eugenic Solution," *Sociological Papers* 3 (1907): 55-80, with "Discussion," 81-104.

It bears noting that there were some few hereditarian psychologists who seem not to have been directly influenced by Galton. The most important of these was the educational psychologist Henry Herbert Goddard, who is noted for introducing intelligence testing to the United States and promoting "the myth of the menace of the feebleminded." Goddard was too deeply involved in the soft-heredity-based program best represented by G. Stanley Hall to have paid much attention to Galton and his eugenics around the turn of the century, and when he did imbibe his Galtonism, in 1909, he got it in predigested and reconstituted form, from Charles B. Davenport.

CHAPTER TWO
GALTON'S EVIDENCE

IN PUBLISHING "Hereditary Talent and Character," his initial statement on eugenics, in *Macmillan's Magazine* Galton faced a markedly liberal audience. His two-part manifesto, published in the June and August 1865 numbers, stood in bold contrast to its contiguous submissions, which included an appreciative review of a book celebrating the Wedgwood family as successful climbers from the artisan to the middle class; an appraisal of the recently-martyred Lincoln as the natural superior of the more nobly-born Washington; and a recognition of "connexions" as a major determinant of modern worldly success.[1] Galton advanced the claim that intelligence and the accompanying qualities required for worldly success are inherited, that eminence runs in prominent families not because of these families' superior social legacy, their relative advantages and connections, but as the natural product of their superior biological heredity. The processes of natural selection illuminated by his half-cousin Charles Darwin had produced, over the course of the centuries, an admirably diligent English race. Just so, Galton promised, could artificial selection mold, in remarkably shorter order, a greatly more intelligent and moral one.[2]

---

[1] Francis Galton, "Hereditary Talent and Character," *Macmillan's Magazine*, "Part I," June 1865, 157-66; "Second Paper," August 1865, 318-27. "Miss Meteyard's Life of Wedgwood," review of *Life of Josiah Wedgwood, from His Private Correspondence and Family Papers; with an Introductory Sketch of the Art of Pottery in England*, by E. Meteyard, *Macmillan's Magazine*, June 1865, 155-57; Edward Dicey, "Lincolniana," *Macmillan's Magazine*, June 1865, 192; as one of the ongoing "Essays at Odd Times," "Of Success in Life," *Macmillan's Magazine*, June 1865, 181-85.

Viewed in retrospect, 1865 turned out to be a momentous year for the study of heredity: Francis Galton first published his claims for what would be eugenics; Gregor Mendel completed his study of sweet peas and read his historic paper; and the American paleontologists later to be termed Neo-Lamarckians initiated their concerted investigation of heredity processes.

[2] Galton's reference to the bred-in diligence of the English people appears in "Hereditary Talent," "Second Paper," 325; his claims for artificial selection are adumbrated in "Part I," 157-58, and spelled out there on 165-66:

> If a twentieth part of the cost and pains were spent in measures for the improvement of the human race that is spent on the improvement of the breed of horses and cattle, what a galaxy of genius might we not create!

Yet in terms of proof provided, Galton's first elaboration of his theories was grossly inadequate. The whole of the evidence for his contention that intelligence is determined by heredity resided in the commonplace that the eminent boast a larger proportion of eminent relatives than others do. Rather than address this phenomenon as something needing to be explained, Galton contented himself merely with documenting it, providing six tallies of eminent figures — two from mixed fields of endeavor, and four from discrete occupational groups — along with their eminent relatives, as derived from various reference works. Certainly he attempted to control for family advantage, the admitted alternative to his own explanation, by including among these groups one survey which treated distinguished figures from the "more open fields" of science and literature. But here he conducted himself curiously. After acknowledging that this inquiry would provide a useful test of his theory, Galton consolidated his litterateurs and scientists, failing to specify what numbers had come from each field and omitting the names of the eminences he had discovered. Strangely, only for this, his most difficult and telling category, did he maintain that his related figures had distinguished themselves in the same fields. More specifically, he made the extraordinary claim that one in every 6 ½ literary or scientific figures of eminence over the past four centuries had had at least one relative of comparable distinction laboring in the same vineyard.[3] Taken by themselves, such numbers could have been impressive, equaling as they did the proportions he had claimed for other fields more clearly susceptible to family

---

. . .Men and women of the present day are, to those we might hope to bring into existence, what the pariah dogs of the streets of an Eastern town are to our own highly-bred varieties.

He also explains on p. 326 of the "Second Paper" that "inferior blood in. . .a family might be eliminated from it in a few generations" of wise breeding. On 319-20 he explains the process whereby a given population could eradicate its "refuse," its "weakly and incapable men," "in a few generations."

[3] Discussion of Galton's biographical material is in "Part I," 159-64. His claim that one in 6½ literary or scientific figures of eminence had at least one relative of comparable eminence in the same profession appears in the same installment on p. 161.

influence. Had Galton not combined his two occupations and had he specified the individuals selected, his argument might have been nicely strengthened.

Yet overall the presentation of his case to this point suffered in ways that would prove characteristic. For almost all of his evidence for eugenics was to be marked by a combination of deeply purposeful planning regarding issues to be addressed and astonishing carelessness in execution. So jarring a mixture of qualities makes it difficult to distinguish tendentiousness from near-pathological inconsistency. As one example that could be checked, there is his use of a reference work entitled *Men of the Time*, which Galton described as containing "an account of the distinguished men of England, the Continent, and America, who are now alive." In actuality the book treated men — and women — from all parts of the world. Galton claimed there to be eighty-five entries under the letter *A*; in reality there were ninety-seven, although there were indeed eighty-five from the geographical areas mentioned by him. Of these eighty-five names, he claimed, "no less than 25 of these, or 1 in 3½, have relatives also in the list. . . ." Yet ten of these twenty-five may have had eminent relatives but none "also in the list," either under the *A*'s or elsewhere, which Galton himself indicated by not counting these relatives. Accordingly these ten should not have qualified by the terms of Galton's own specifications. Here he seems carelessly inconsistent, at best. One other of Galton's entries of the same sort he treated differently, by counting a (dead) brother as appearing in the book's list although he had not. Here he seems doubly inconsistent and possibly dishonest. Four more names he slipped onto his own list despite the fact that they appeared nowhere in the book, which seems entirely deceitful on his part. Furthermore, he claimed Queen Victoria's son Prince Albert (Albert Edward, the Prince of Wales, later to reign as Edward VII) for having an eminent brother, Prince Alfred, yet he did not enter Alfred despite the fact that he too had appeared in the book, under the *A*'s. Seemingly, his apparent reason for not entering Alfred should have disqualified Albert: both, being in their early twenties, had not as yet distinguished themselves as eminent except as royalty. Moreover, Galton had earlier imposed upon himself a restriction against including royalty, as necessarily eminent. In sum, sixteen of Galton's twenty-five names (64 percent) were improperly — if not spuriously — claimed, and rather than 1 in 3 ½, his ratio had been 1 in 9 $^4/_9$ (9 in 85), or 1 in 10$^7/_9$ (9 in 97).[4] Were these errors the product of mere carelessness or deceptiveness? If the latter, Galton's could only have been a most reckless deceptiveness, inasmuch as his source had been readily available to anyone who cared to evaluate his performance. Examples of so heedless a straining after victory could be presented at length. "Hereditary Talent and Character" succeeded only as a position paper, an affirmation of first principles affording no real proof.

---

[4] The treatment of the material from *Men of the Time* appears in "Part I," 160-61. Galton's reference to the author of the work as "Walford" may have been an error: no author is listed for any of the work's various editions. Examination and comparison of these editions establishes that Galton had to have been consulting the series' sixth edition, which is *Men of the Time: A Biographical Dictionary of Eminent Living Characters of Both Sexes* (London: George Routledge and Sons, 1865). Listings of names beginning with the letter A appear on pp. 1-32. The ten individuals appearing in the book who did not have relatives "also in the [book's] list" were Sir William A'Beckett, Jean Victor Adam, Charles Francis Adams, Joseph Addison Alexander, Sir Archibald Alison, Etienne Arago, Pericles Argyropoulo, Nicolas Aristarchi, Matthew Arnold, and Adolf-Iwar Arwidson. The brother whom Galton counts although he had not appeared in the book's list was James Waddell Alexander. The four names which Galton includes despite their absence from the list are Sir Charles Aldis, Charles J. B. Aldis, André Marie Ampère, and Lord Ashburton. Such smugglings-in seem especially egregious in view of the assurance he had provided on p. 159 that "it did not. . .much matter whose biography I adopted. . .so long as I determined to abide stedfastly [sic] within its limits, without yielding to the temptation of supplying obvious omissions, in a way favourable to any provisional theory."

Note also, as merely an interesting sidelight, that effectively Edward VII would repay Galton for having identified him as eminent so many years before when the monarch knighted him in 1909.

Finally, note that Ruth Schwartz Cowan discusses various deficiencies in Galton's presentation of evidence in "Hereditary Talent and Character," in "Sir Francis Galton and the Study of Heredity in the Nineteenth Century," Ph.D diss., Johns Hopkins University, 1969, 5-10, and also in a condensation of the material in her dissertation, in Cowan, "Nature and Nurture: The Interplay of Biology and Politics in The Work of Francis Galton," *Studies in the History of Biology* 1 (1977): 134-36.

Galton's next attempt at establishing the overwhelming heritability of intelligence came four years later, in another *Macmillan's* piece. In March 1869, he published "Hereditary Genius: The Judges of England between 1660 and 1865," an article designed as a forerunner to a book in preparation, some of whose proof-sheets he claimed to have before him as he wrote.[5] The corroboratory plan of this article was identical to that of its predecessor: Galton had studied a listing of English judges of the period specified, and isolated all with eminent relatives. Of the 286 judges in his time period, he claimed to have found 133 with relatives of eminence sufficient for inclusion. Here again, however, he failed to name most of the judges and relatives selected, pleading limitations of space. Moreover, although the article steered clear, by definition, of literary and scientific figures, and thereby of a more meaningful test of his heredity propositions, it did reiterate more forcefully an assertion advanced in the previous discussion of such classes. As one of the propositions which, he wrote, should obtain if his general theory were correct, Galton again expressed the idea that "the

---

[5] Galton, "Hereditary Genius," 424-31. Curiously enough, this article has never been listed in any of the bibliographies of Galton's published work, although he makes direct reference to it on the errata page opposite the first page of *Hereditary Genius*. "I am glad of this opportunity to correct a mistake in an article I published this spring," he writes, "in *Macmillan's Magazine*, on the 'English Judges.'" One possible explanation for this universal oversight might be that all commentators on Galton who have noticed these lines have assumed (somehow) that he was referring to his treatment of Lord Chancellors in "Hereditary Talent and Character," some four years earlier. At any rate, "Hereditary Genius" merits critical scrutiny, however belated, especially inasmuch as it opens with Galton's stipulation of five "conditions" which should "necessarily be found to exist . . .if genius be hereditary in the same way that physical strength or feature is hereditary. . . ." (424). The third and fifth of these conditions appear nowhere else, apparently because Galton was forced to abandon them. The former is discussed in the text; the latter is that "The appearance of the man of hugest ability in a family should not be an abrupt and isolated phenomenon, but his ability should be built up so to speak, by degree, in his ancestry; and conversely, it should disperse itself by degrees in his descendants." For the reference to this article, see Galton, *Hereditary Genius: An Inquiry into Its Laws and Consequence* (London: Macmillan and Co., 1869), viii. Note that the same reference to "Hereditary Genius" appeared in the 1892 edition of the book.

peculiar type of ability ought to be transmitted," a contention he had largely implied before.[6]

The book Galton promised came out in November of the same year. *Hereditary Genius: An Inquiry into Its Laws and Consequences* cut a distinctively modern figure, seemed poised, to many readers, on the frontiers of science. In an early chapter, in treating the probable bell-shaped distribution of ability in general populations, Galton emphasized the normal law of error, describing it as deserving "to be far better known to statisticians than it appears to be." In the book's closing chapter he demonstrated an acquaintance with Darwin's freshly-minted provisional theory of pangenesis.[7] Much of this was window-dressing. Between these ornaments the study evidenced both a narrowing and a retrogression in Galton's campaign to secure his heredity claim. While *Hereditary Genius* widened the investigation in terms of occupations studied, increasing these to twelve, it represented no advance in evidentiary strategy. Here again Galton's entire proof of the prevalence of heredity in intelligence determination consisted of the demonstration that the eminent boast a higher proportion of eminent relatives than the masses. Moreover, the book constituted a bitter reckoning for several important claims previously tendered on faith. Key representations, once forced into the open, shrank dramatically. The section discussing the English judges treated in the previous publication, one now naming these judges and their relatives of eminence, showed a silent reduction in number from 133 to 109 out of 286.[8] Something in the process of specifying his choices had

---

[6] "Hereditary Genius," 426 for the numerical claim; 424 for the third "condition."

[7] For the claim that *Hereditary Genius* was published in November 1869, see Karl Pearson, *The Life, Letters and Labours of Francis Galton*, vol. 2 (Cambridge: Cambridge University Press, 1924), 88n., where Pearson quotes from Galton's wife's record of noteworthy events for that year. See *Hereditary Genius*, 26, for Galton's remark about the normal law; 26-36 for his elaboration and application of that device, which is further discussed in an appendix on 377-83; 363-73 for his discussion of pangenesis.

[8] Ibid., 55-315 for the discussion of the biographical information for the twelve occupational groups; 58 for the revised numerical claim regarding the judges and

constrained Galton to throw out nearly one in five (18 percent) of those previously claimed.

His treatment of his crucial Literary Men suffered far more. Here Galton was forced to abandon all pretense of limiting himself to independent listings. Instead he ranged freely through four centuries to locate forty-three names for discussion (which he somehow counted as thirty-seven, thereby boosting his ratio of relatives-per-entrant) and to assemble his customary list of eminences with no apparent relatives of eminence. This latter list, which omitted mention of such figures as Shakespeare, Goethe, Milton, Molière and Schiller, could, of course, have run on indefinitely. Instead it stopped at nineteen. Curiously, after this tortured exercise Galton elected to conclude that "the general result of my inquiries is such to convince me that more than one-half of the great literary men have had kinsmen of great ability." Yet instead of one in 6 ½ of all literary figures of distinction showing at least one close relative of equal eminence in the same field (his claim in "Hereditary Talent and Character"), Galton was able to supply only three such cases: the symbiotically reclusive Brontës; the Scaligers, father and son; and the distantly-connected Dryden and his aunt's grandson Swift. (In his Poets section, which searched through all of history, only Dryden and Swift satisfied this criterion, whom he pressed into service again.) Perhaps it was because of this dismal showing in so critical an area that sometime between writing "Hereditary Genius," the article, and publishing the book of the same name Galton quietly abandoned his claim that specific talents are inherited. Yet still he seems to have wanted to have things both ways. Even after being forced to forage in fields more open to family influence for relatives for his writers, he persisted in assuring readers that "we may rest satisfied that an analysis of kinsfolk shows literary genius to be fully as hereditary as any other kind of ability we have hitherto discussed."[9]

---

their eminent relations.

[9] See ibid., 167-91 for Galton's overall discussion of his Literary Men; 172 for his specification of thirty-seven as "the total number of names included in my list of kinships. . . ," for his list of nineteen eminences without eminent relatives, and for

Moreover, even with his newly-expanded construction of ancestral transmission victory proved elusive. In order to fill his literary list he was compelled to relax greatly, indeed to suspend without announcing as much, his standard of eminence. He qualified Chateaubriand for having an apparently unpublished sister who "resembled him in genius and disposition"; Maria Edgeworth for an undistinguished father whom she much assisted in his writings; and the scholar and critic Richard Porson for a father who had been a weaver and parish clerk, a mother who was a "housemaid at the clergyman's, who read his books on the sly," a brother who had died young but was reputed to be his equal, and a sister said to possess "the wonderful Porson memory." He claimed Madame de Sévigné for having a son without "sufficient perseverance to succeed in anything" and a nephew who wrote good letters but was exiled for being "ill-natured" and "caustic"; and Le Sage for two of his three sons: a short-lived abbé turned stage comedian, and a "jolly" canon who "enjoyed life, and loved theatricals, and would have made an excellent comedian." He included Washington Irving despite the fact that he had had no relatives whom Galton could discuss without enclosing their relationship symbol in (disqualifying) brackets.[10] Other categories presented similar problems. In a subsequent section, in an attempt to show that exceptional physical qualities are as heritable as mental ones, he treated "Wrestlers of the North Country." Here he ignored the advice of his chief informant that champions typically become bar owners and draw their sons off in this new direction, dispensed with his practice of compiling some listing of eminent figures with no eminent relatives, and, after much additional work on his informant's part, set forth a list of twenty wrestlers (whom he counted as eighteen) with variously-successful wrestler-relatives as proof—notwithstanding the significant majority lacking such

---

his remark that "one-half of the great literary men have had kinsmen of great ability"; and 171 for his statement that "literary genius" is "fully as hereditary as any other kind of ability...hitherto discussed."

[10] See *Hereditary Genius*, 174-75 for Chateaubriand, 175 for Edgeworth, 180 for Porson, 183 for de Sévigné, 181 for Le Sage, and 178 for Irving.

relatives—of his basic contention that great ability is generally a product of heredity.[11] Once again, such examples could be extended almost indefinitely. Considered as a vindication of its author's claims for the predominance of heredity and the insignificance of family advantage, *Hereditary Genius* was an abject disaster.

There are at least three things that might be pointed out in Galton's performance in the book. The first is the ostensible superfluity of the argument he presented there. Clearly he was not required to engage in any of his attempted evidence-gathering if all he had wanted to contend was that some form of artificial selection could improve humanity in almost any way chosen. The astonishing successes scored by plant and animal breeders had already advanced that argument for him, for all the world to see. Clearly he was arguing something more: first, certainly, that individual intelligence is maximally limited, differential (individually variegated in its limits), and heritable; and second, slightly less obviously, that dependably good intelligence inheres to the superior social classes, that a natural biological sifting had already occurred, with the cream settling naturally, effortlessly, at the top. This idea that Galton was so laboriously (and tacitly) demonstrating, the bolder of his English and American heirs would term, in occasional bursts of candor, his "superiority doctrine."[12]

---

[11] For the discussion of Wrestlers of the North County, see ibid., 312-15; for Galton's enumeration of them, see 312. For correspondence with Galton of Robert Spence Watson, his chief investigator and informant for the wrestlers and also for his Oarsmen, see Watson to Galton, 3 December 1867; 16 January, 30 January, 14 November, and 22 December 1868; Folder 120/2, Francis Galton Collection, Manuscripts and Rare Books Room, Science Library, University College London, London, England, hereafter referred to as Galton Collection.

It might also be noted in this context that Galton appears also to have considered the Northumberland sport of clear-water trout-fishing for inclusion in *Hereditary Genius*, after discovering that championships were often won by sons of previous champions. But after Watson recounted anecdotes of these fathers having pitched their sons bodily into streams for small variations of form, Galton terminated this inquiry. See especially Watson to Galton, 25 November 1868, Folder 120/2, Galton Collection.

[12] William McDougall, when head of Harvard University's psychology division, quoted one of his books in defining the " 'superiority doctrine' " as the contention "'. . .that the upper social strata, as compared with the lower, contain a larger

The second thing to be seen throughout Galton's gathering of evidence for eugenics is that it was something other than scientific inquiry, even by the standards of his day. What Galton proffered as science the nineteenth century still termed *eristics*, a highly useful term that fell out of the English language sometime near the end of that century. Aristotle best articulated the operative dichotomy: whereas dialectics is a process devoted to establishing truth, an open contest with no predetermined victor, eristics is a process directed solely toward victory, a means of establishing a previously chosen conclusion no matter what new truths might be uncovered along the way.[13] Thus science might be classed as a dialectic process and such activities as politics and sales as eristic ones. Certainly, scientific inquiry, being a human enterprise, often admits of some admixture of eristics, but the science so prominent in Galton's eugenics to this

---

proportion of persons of superior natural endowments.'" See "Correspondence," *New Republic,* 27 June 1923, 125.

[13] Aristotle, *Sophistici Elenchi* 11.171^b. 20-27, as in *Aristotle on Fallacies Or the Sophistici Elenchi,* trans. Edward Poste (London: Macmillan and Co., 1866), 34-39. For discussions of this distinction, see G. E. L. Owen, "Dialectic and Eristic in the Treatment of the Forms," in ed. G. E. L. Owen, *Aristotle on Dialectic — The Topics* (Oxford: Oxford University Press, 1968), 103-07; J. D. G. Evans, *Aristotle's Concept of Dialectic* (Cambridge: Cambridge University Press, 1977), 8, 14, 76n., 91.

For an indication of the declared standards of the period, note Thomas Huxley's statement regarding the nature of his commitment to Darwin's theory of natural selection:

> . . .you must understand that. . .I accept it provisionally, in exactly the same way that I accept any other hypothesis. Men of science do not pledge themselves to creeds; they are bound by articles of no sort; there is not a single belief that it is not a bounded duty with them to hold with a light hand and to part with it, cheerfully, the moment it is really proved to be contrary to any fact, great or small.

See Huxley, *On the Origin of Species: Or, the Causes of the Phenomena of Organic Nature. A Course of Six Lectures of Working Men* (New York: D. Appleton & Co., 1863), 145.

For a useful discussion of calculating misrepresentation in science throughout history, see William Broad and Nicholas Wade, *Betrayers of the Truth* (New York: Simon and Schuster, 1982), especially the appendix, "Known or Suspected Cases of Scientific Fraud," 225-32.

point was primarily and fundamentally eristic. Clearly Galton was promoting a program of some sort, with a much-constrained science entirely in its service.

The third thing that might at least be inferred throughout *Hereditary Genius* is Galton's own awareness of the insufficiency of his argument to attain the victory he kept so strenuously claiming and seeking. This sensibility stands out from the first, in his preface, where he assured that "the foundation for my theories is broader than appears in the book," and requested that the reader bear as much in mind "as a partial justification if I have occasionally been betrayed into speaking somewhat more confidently than the evidence I have adduced would warrant." Similarly, in his introduction he pleaded with the reader to "make allowance for a large and somewhat important class of omissions I have felt myself compelled to make when treating of the eminent men of modern days": omissions of the "names of their relations in contemporary life who are not recognized as public characters, although their abilities may be highly appreciated in private life." Galton claimed to have been prevented from citing these figures "by a sense of decorum," whereas in actuality he had been circumscribed by his own rules—he had claimed, after all, to be able to show that the eminent have an unusually large proportion of relatives who also are eminent. His reason for introducing such an issue becomes apparent when he asks his readers to bear these decorous omissions in mind so as to more than counterbalance "such errors as I may and must have made which give a fictitious support to my arguments. . . ." Galton's concern for the actual strength of his case might also be heard in his reference in the book's introduction to the laurels already conferred on his argument (as expounded in the magazine pieces) by "many of the highest authorities on heredity" (none of whom named) and by his famous cousin.[14] (The year before, in his *Variation of Animals and Plants Under Domestication*, Charles Darwin had indeed declared Galton to have demonstrated that "complex mental attributes, on which genius and talent depend, are inherited," but this support appears to have been *pro forma*.)[15]

[14] *Hereditary Genius*, v, vi; 3, 3, 3, 4; 2.

[15] *The Variation of Animals and Plants Under Domestication*, 2 vols. (London: John

Perhaps the most significant indications of Galton's consciousness of the insufficiency of his argument are provided by his employment of entirely untestable claims such as the one that a "considerable proportion" of the four hundred greatest geniuses in all of world history (all of whom were unnamed) "will be found to be interrelated"; the examples of heedless and sometimes deceitful-seeming straining after victory previously presented; and, most important of all, his careful refusal ever to specify in *Hereditary*

Murray, 1868), vol. 2, p. 7.

In his *The Descent of Man and Selection in Relation to Sex*, 2 vols. (New York: D. Appleton & Co., 1871), Darwin paid further homage to *Hereditary Genius*. In vol. 1, 106-07, he remarked that ". . .we now know through the admirable labors of Mr. Galton that genius, which implies a wonderfully complex combination of high faculties, tends to be inherited," and in vol. 1, 161n., he termed the book "his great work." All the same, although Darwin went on to issue the observation increasingly heard at the time, that medical progress was resulting in the rescue of weakly constituted types who would have perished in earlier times—a course which must be "highly injurious to the race of man" — still he argued that checking "our sympathy" for these unfortunates would necessarily result in "deterioration in the noblest part of our nature" (vol. 1, 161-62). Negative eugenics (as it would later be termed) should be implemented only on a voluntary basis. Moreover, in keeping with his doctrine of natural selection, and because intelligence constitutes an obvious competitive advantage, Darwin argued that "in civilized nations there will be some tendency to an increase both in the number and in the standard of the intellectually able" (vol. 1, 164-65). Not only this, but the morally vicious are steadily eliminated through the prevention of their reproduction by their intemperance, profligacy promoting disease, and incarceration (vol. 1, 165-66). The tendency which W. R. Greg and others were reporting, of superior strains being much outbred by inferior ones, was being compensated for by such factors as the latter class's higher mortality rate (vol. 1, 168-69). In sum, *Descent of Man* showed that Darwin, though cognizant of all the doom-shouting of his time, was not persuaded that humanity was in danger, nor that a program such as Galton's was indicated. For that matter, the book shows Darwin to have been aware of the largely negative reviews that Galton's book had received (see Footnote 19), and suggests that in praising the work without actually discussing or defending it textually, he may largely have been concerned with counterbalancing such criticism generally, in the interest of family loyalty.

For an extended consideration of Darwin's views on the social evolution of man, and of the influence exercised by other writers including Galton upon these views, cf. John C. Greene, "Darwin as a Social Evolutionist," *Journal of the History of Biology* 10 (Spring 1977): 1-27.

Genius any standard of eminence whatever for the crucial relatives of his figures of eminence.[16]

The more important reviews of *Hereditary Genius* proving to be generally negative, Galton was at work on a "supplementing" of the book in 1872, when the Swiss naturalist Alphonse de Candolle, one of the figures cited in Galton's "Men of Science" section, published a comparative study of the production, by various European nations, of scientists over a 119-year period.[17] De Candolle's book challenged

---

[16] See *Hereditary Genius*, 2-3, for the claim regarding the four hundred greatest geniuses.

[17] In regard to the critical reception of *Hereditary Genius*, Louisa Galton noted in her annual record that the book though "not well received," was "liked by Darwin and men of note." See Pearson, *Life of Galton*, vol. 2, 88n. A sampling of the British reviews indicates that the first part of her entry was accurate. "Hereditary Genius," *Saturday Review*, 25 December 1869, 832-33, constituted a sedately dismissive appraisal:

> . . .Mr. Galton has secured a draught of evidence which we can but characterize as largely mediocre and such as points with infinitely greater truth to the influence of generally diffused and high-pitched culture than to anything of the nature of inherent genius following upon a strain of blood.

". . .Mr. Galton's elaborate figures may be as easily turned to support the very opposite conclusion. . .the long array of names and figures which are made to prop up the the hypothesis of hereditary genius, however interesting as bits of biography, seems to us logically worth nothing" (833, 833). "History and Biography," *Westminster and Foreign Quarterly Review* 93 (1 January 1870), 300-02, was more positive, according his "bold speculations" and "daring and suggestive inquiry" general success: "The instances in which a plurality of capable descendants or kinsmen are found in the same family are too numerous and too marked to allow of any other explanation than that on which Mr. Galton insists" (300-01). "Critical Notices: Some Books of the Month," *Fortnightly Review*, 1 February 1870, 255, provided a neutral and thereby relatively positive-seeming abstract of the book's argument and contents. Alfred R[ussel]. Wallace, "Hereditary Genius," *Nature* 1 (17 March 1870): 501-03, comprised a reservedly positive review, perhaps reflecting Wallace's disinclination to derogate the efforts of Darwin's cousin, and the reluctance of Norman Lockyer, the publisher of the newborn *Nature*, to issue a negative account of so strenuous a work by one of his friends and former partners at *The Reader*. (Certainly *Nature* took Galton's side in all conflicts thereafter—even the one between the forces of biometrics and Mendelism, as long as this support remained possible.) Frances Power Cobbe, "Hereditary Piety,"

Galton's hereditarian explanation of eminence explicitly, and seems to have so impressed Darwin that in January of 1873 he warned Galton that "we ought both to shudder using so freely the word 'Nature' after what De Candolle has said."[18] Galton responded with

---

Theological Review: A Journal of Religious Thought and Life (29 April 1870): 211-34—which was reprinted in her *Darwinism in Morals, and Other Essays* (London: Williams and Norgate, 1872), 35-63—was resolutely positive, albeit largely concerned with Galton's remarks on his Divines. George Harris, "Hereditary Genius," *Journal of Anthropology* 1 (July 1870): 56-65, expressed a generally skeptical sense of Galton's claims and achievement in the book, but recognized the importance of the study of heredity as a new and promising field. The most scathingly negative of these reviews, [Herman Merivale (according to Cowan, "Nature and Nurture," 201n.),] "Hereditary Genius: An Inquiry into its Laws and Consequences," *Edinburgh Review*, July 1870, 100-25, accused Galton of overstating his case, misusing evidence, and engaging in fuzzy thinking, imprecise terminology, and (the implication seems clear) special pleading. [Rev. F. W. Farrar,] "Hereditary Genius," *Fraser's Magazine*, August 1870, 251-65, started out appearing quite impressed by Galton's honesty and industry, then turned on him in fairly devastating fashion as well.

   The American reviews, which probably mattered far less to Galton, were generally positive. See "Hereditary Genius," *Appleton's Journal of Popular Literature, Science, and Art*, 19 February 1870, 217-18; review of *Hereditary Genius*, Galaxy, March 1870, 424; review of *Hereditary Genius, Harper's New Monthly Magazine*, May 1870, 928-29; "Hereditary Genius," *Atlantic Monthly*, June 1870, 753-56. For an emphatically negative American review, see [Orestes A. Brownson,] "Hereditary Genius," *Catholic World*, September 1870, 721-32. Brownson opened by assailing Darwinism and modern science generally but closed in on Galton, explaining that "we refuse. . .to accept Mr. Galton's hypothesis that genius is hereditary, because the facts he adduces are not all the facts in the case, because there are facts which are not consistent with it, and because he does not show and cannot show that it is the only hypothesis possible for the explanation even of the facts which he alleges" (727).

   See also Emel Aileen Gökyigit, "The Reception of Francis Galton's *Hereditary Genius* in the Victorian Periodical Press," *Journal of the History of Biology* 27 (1994): 215-40.

   Galton's statement that he was working at "supplementing" *Hereditary Genius* in 1872 appears in his *English Men of Science: Their Nature and Nurture* (London: Macmillan, 1874), v.

   De Candolle's study was *Histoire des Sciences et des Savants depuis Deux Siècles suivie D'Autres Études sur des Sujets Scientifiques en Particulier sur la Sélection dans L'Espèce Humaine* (Geneva: H. Georg, Libraire-Éditeur, 1873).

[18]Darwin to Galton, 4 January 1873, reproduced in Pearson, *Life of Galton*, vol. 2, 176. De Candolle's discussion of heredity appears in *Histoire*, 92-96, 308-36.

a tart review in 1873, and in 1874 with another article and a book of his own.[19] Essentially a study of self-report questionnaires returned by English scientists requested to describe the influences on their choice of career, *English Men of Science* opened with Galton's characteristic assurance that "one effect of the evidence here collected will be to strengthen the utmost claims ever made for the recognition of the importance of hereditary influence." But the 100 responses that Galton had selected from the 180 returned proved quite mixed, and Galton ended up effectively conceding the force of de Candolle's environmentalist argument with his acknowledgment that social forces had been retarding the production of English scientists.[20]

Two other ostensible sources of proof often cited by American hereditarians are his discussion of the popes' nephews in *Hereditary Genius* (whom Galton had used to show the inefficacy of family influence in securing eminence), and his pioneering studies of twins, first published in 1875.[21] Such appeals to Galton's authority often

[19]Galton, "On the Causes Which Operate to Create Scientific Men," *Fortnightly Review*, 1 March 1873, 345-51; "On Men of Science, Their Nature and Nurture," *Proceedings of the Royal Institution* 7 (1874): 227-36; *English Men of Science*. For an excellent discussion of Galton's and de Candolle's interaction regarding de Candolle's book, and Galton's response to de Candolle's criticisms, see Raymond E. Fancher, "Alphonse de Candolle, Francis Galton, and the Early History of the Nature-Nurture Controversy," *Journal of the History of the Behavioral Sciences* 19 (October 1983): 341-52. See also, for an extensive appraisal of de Candolle's book, W.H. Larrabee, "De Candolle on the Production of Men of Science," *Popular Science Monthly* 29 (May 1886): 34-46. Pearson reproduces a number of Galton's and de Candolle's letters to each other during the period of December 1872 to May 1876, and November 1879 to September 1885, in *Life of Galton*, vol. 2, 131-49, 204-10, and more for the period of October 1885 to June 1890, in Pearson, *The Life, Letters and Labours of Francis Galton*, vol. 3B (Cambridge: Cambridge University Press, 1930), 474, 476-81, 483. He also provides a useful discussion of their influence upon each other in his *Life of Galton*, vol. 2, 131-56. Helpful discussions of the context of *English Men of Science* also appear in Ruth Schwartz Cowan's introduction to a recent edition of that work (London: Frank Cass, 1970), and in Victor L. Hilts, *A Guide to Francis Galton's English Men of Science*, Transactions of the American Philosophical Society n.s. 65, Part 5 (1975): 1-85.

[20]*English Men of Science*, vii, 258-60.

[21]*Hereditary Genius*, 42; "The History of Twins as a Criterion of the Relative Powers

celebrated the statistical rigor of these two treatments. In reality Galton's discussion of the first group had consisted entirely of the offhanded assertion that, from what he had seen, not having "worked up the kinships of the Italians with any especial care," these nephews, "whose advancement has been due to nepotism, are curiously undistinguished."[22] His study of twins was composed of anecdotal descriptions supplied by lay observers, usually parents. Neither study had involved statistical analysis. That Galton's American heirs could hardly have believed their representations of him as having established the all-importance of heredity to intelligence can best be seen in their abstemious disinclination ever to stoop to any real discussion of his alleged proofs. They knew better than that.

---

of Nature and Nurture," *Fraser's Magazine* (November 1875), 566-76; "Short Notes on Heredity &c, in Twins," *Journal of the Anthropological Institute of Great Britain and Ireland* 5 (January 1876): 324-29; "The History of Twins as a Criterion of the Relative Powers of Nature and Nurture," *Journal of the Anthropological Institute of Great Britain and Ireland* 5 (January 1876): 391-406.

[22]*Hereditary Genius*, 42. Note that in 1903 Thorndike termed Galton's twin study "the best and indeed the only quantitative study of the potency of original nature. . . .," and that he also observed that "Galton demonstrates that the adopted sons of popes do not approach equality in eminence with the real sons of gifted men." See Thorndike, *Educational Psychology* (1903), 53 for both.

For hereditarian propagandists' uses of these materials, see, for the popes' nephews, Stoddard, *Revolt Against Civilization*, 49-50; Paul Popenoe and Roswell Hill Johnson, *Applied Eugenics* (New York: Macmillan and Co., 1918), 16, where reference is made to "statistics" on the matter; and Albert Edward Wiggam, *The Fruit of the Family Tree* (Garden City, NY: Garden City Publishing Co., 1924), 214, where the claim is made that "Galton showed that, notwithstanding the splendid environment furnished these adopted sons by the popes, they did not rise to eminence as often as did the actual sons of distinguished men." For the twin study, see ibid., 125-28, esp. 125, where particular notice is taken of the "careful methods" employed by Galton in this study.

CHAPTER THREE
BIOGRAPHICAL FACTORS

IF FRANCIS GALTON failed to prove that heredity is the determining factor in the individual's intellectual development, why then was he so unmistakably important to the American psychologists who represented themselves as his heirs? What did he transmit to them? Clearly the point to which they were particularly receptive and which exerted the most influence was Galton's spirit and underlying purpose, which they perceived as animating his formulation of eugenics. Reconstructing their sense of Galton and his invention requires two activities: certainly that we attempt to situate him contextually as they seem to have done, but in addition, we should examine various features of his life up to the creation of eugenics. This biographical consideration, it should be noted, is one which only relatively recent writers have been able to effect. The American hereditarian psychologists were incapable of conducting such an inquiry until 1924 because very little of the necessary information was available to them before then. Galton himself, in the autobiography he published in 1908, was characteristically reticent with regard to (to use his term) the personal "pre-efficients" of eugenics.[1] The obituaries published after his death in 1911 were little more helpful, and it was not until the publication of the second volume of Karl Pearson's massive biography of Galton that speculation could begin as to psychohistorical impetuses.[2] (Pearson's first volume,

---

[1] Galton, *Memories of My Life* (London: Methuen & Co., 1908). In June 1908 Galton read a paper on eugenics before a small group associated with the newly-formed Eugenics Education Society wherein he explained that at some point his experiences as an undergraduate at Cambridge had inspired an interest in the inheritance of mental and physical abilities. See "Eugenics," *Westminster Gazette*, 26 June 1908, 1-2, and D. W. Forrest, *Francis Galton: The Life and Work of a Victorian Genius* (London: Paul Elek, 1974), 276.

[2] [Karl Pearson,] "Francis Galton. February 16, 1822 - January 17, 1911," *Nature* 85 (2 February 1911): 440-45; Montague Crackanthorpe, "Sir Francis Galton, F. R. S., A Memoir," *Eugenics Review* 3 (April 1911): 1-9; Francis Darwin, "Francis Galton, 1822-1911," *Eugenics Review* 6 (April 1914): 1-16; J. Arthur Thomson, "Sir Francis Galton," *Sociological Review* 4 (April 1911): 141-42; [George Howard Darwin,] "Galton, Sir Francis," *Dictionary of National Biography Supplement, January 1901-December 1911*, vol. 1, Abbey-Eyre, ed. Sir Sidney Lee (London: Oxford University

published in 1914, carried Galton only to 1854.) All the same, even
if the following biographical conceptualization was impossible for
the earlier American hereditarian psychologists, it can variously
offset, augment, and introduce the view which they seem to have
taken of contextual factors determining their mentor's development
of eugenics.

It was the initial irony of a life to be marked by many that
Francis Galton was born—on February 16, 1822, one mile outside
Birmingham—at "the Larches," the former home of Joseph Priestley,
the great pioneer of chemistry and embattled early champion of
democratization in England.[3] (In 1791 a mob exercised over French
republicanism had destroyed Priestley's laboratory, books, manuscripts, and all but one room of the residence, which a subsequent,
intervening owner rebuilt.)[4] Irony number two commenced
immediately at Galton's birth, with the ascertainment of his gender.
For his parents had already resolved that, if their child were male

---

Press, 1912), 70-73.

For other useful appraisals of Galton (in addition to the biographies and bio-
graphical treatments to be cited presently), see Arthur Keith, "Galton's Place
Among Anthropologists," *Eugenics Review* 12 (April 1920): 14-28; Karl Pearson,
Francis Galton 1822-1922: A Centenary Appreciation (London: Cambridge University
Press, 1922); Eliot Slater, "Galton's Heritage," *Eugenics Review* 52 (July 1960): 91-
103; Cyril Burt, "Francis Galton and His Contributions to Psychology," *British
Journal of Social Psychology* 15 (May 1962): 1-49; Norman T. Gridgeman, "Galton,
Francis," *Dictionary of Scientific Biography*, vol. 5, Fisher- Haberlandt, ed. Charles C.
Gillispie (New York: Scribner's, 1972), 265-67; and Solomon Diamond, "Francis
Galton and American Psychology," eds. R. W. Rieber and Kurt Salzinger, *Psychol-
ogy: Theoretical-Historical Perspectives* (New York: Academic Press, 1980), 43-55.

[3] For information on Galton's birthplace, see Forrest, *Francis Galton*, 4, and Pearson,
The Life, Letters and Labours of Francis Galton, vol. 1 (Cambridge: Cambridge
University Press, 1914): 51, 62, 75, 241.

[4] For extensive information on the circumstances, particulars and aftermath of this
incident, see [Alexander Golden and Philip Joseph Hartog,] "Priestley, Joseph,
LL.D.," *Dictionary of National Biography*, vol. 16, Pocock-Robins, eds. Sir Leslie
Stephen and Sir Sidney Lee (London: Oxford University Press, 1917), 363-64. This
account makes mention of the fact that Galton's paternal grandfather had been one
of Priestley's principal patrons.

and it proved at all possible, this final son — who would figure as history's foremost proponent of the proposition that ability is born and not made — would be made into a genius, or, perhaps more properly put, helped to make the most of the genius he might be inheriting. Not only was he born into a wealthy family replete with six significantly older siblings contending for the privilege of nurturing him, he was born to a daughter of one of the seemingly epochal figures of the day just past, the physician, poet, and evolutionist-of-sorts, Erasmus Darwin. That such a heritage mattered deeply to Galton's mother is obvious; he emphasized in a letter describing (to de Candolle) the influences on his early development that she "had never wearied of talking about" his illustrious grandfather.[5] Moreover, as Cyril Burt once explained, the Galtons, whom his father knew as their physician, constituted a wealthy commercial family intent on moving up socially.[6] Such a process typically involved the second or third generations'

---

[5] The letter to de Candolle is reproduced in Pearson, *Life of Galton*, vol. 2, 206, and was written on June 5, 1882. Information on Galton's early life and on his parents' plans for him is provided in Pearson, *Life of Galton*, vol. 1, 62-91; Forrest, *Francis Galton*, 1-26; and C. P. Blacker, *Eugenics: Galton and After* (Cambridge, MA: Harvard University Press, 1952), 19-23. Galton's upbringing and early education are insightfully discussed, and contrasted with those of John Stuart Mill, by Raymond E. Fancher in *The Intelligence Men: Makers of the IQ Controversy* (New York: Norton, 1985), 18-26.

A further indication of Galton's mother's conception of her last-born child might be seen in the opening lines of a biographical sketch she composed of him on the occasion of his being sent to France at age 8 to develop a facility in French: "Francis Galton Son of Sam'l Tertius [&] Violetta Galton [&] Grandson of Erasmus Darwin. M.D. F.R.S. Author of Zoonomia, Phytologia, Botanic Garden [&c] [&c] [&c] [&c] — was born the 16th February 1822. . . ." See Pearson, *Life of Galton*, vol. 1, Plate XL, opposite p. 62.

[6] See Burt, "Francis Galton," 14n. Burt claimed to have been attracted to psychology as a profession through his early acquaintance with Galton, and appreciation of his greatness. For information on Burt's beginnings, see L. S. Hearnshaw, "Burt, Sir Cyril Lodowic," *Dictionary of National Biography 1971-1980*, eds. Lord Blake and C. S. Nicholls (Oxford: Oxford University Press, 1986), 111-12.

It should be noted that Burt's father served as a physician to the Galtons in Warwickshire, beginning in the 1890s, and thus when Francis Galton was in his 70s and Cyril Burt was in his early teens.

distancing themselves increasingly from their trade origins and rising, by means of a university education and the accompanying "connexions," to landed-gentry niches.[7] As his father's family had only recently converted from Quakerism to the Church of England, and his older brothers lacked any interest in academic pursuits, young Francis, should he prove willing and able, would be the first Galton to make this instrumental ascent.[8]

Perhaps, given his remarkable upbringing, it could be expected that if he had anything better than a defective intellect, young Francis would appear exceptionally able, and that willingness would be a natural by-product of such a founding self-definition so insistently assigned. His twelve-year-old sister Adèle, an invalid, had his cradle moved into her room, and taught herself the rudiments of French and Latin so as to stay ahead of her infant charge. Galton was pointing out the letters of the alphabet before he could speak; by age two he was reading books and printing his name; by four he was studying Latin, French, and multiplication.

---

[7]See William O. Aydelotte, "Patterns of National Development: Introduction," in eds. Philip Appleman, William A. Madden, and Michael Wolff, *1859: Entering an Age of Crisis* (Bloomington: Indiana University Press, 1959): 115-30, especially p. 118n.: "[For "eminent business and professional men,"] sending their sons to Public Schools hastened the process of social amalgamation and thus helped to give the second generation of the new class a stamp which made it indistinguishable in externals from the old."

[8] Forrest explains that Galton's father converted to the Church of England after the death of one of their infant children in 1817, five years before Galton's birth. See *Francis Galton*, 4. Galton once wrote to Pearson that his father had been "turned out (automatically) for not marrying his own sect," which would mean that this ejection occurred in 1807. See Galton to Pearson, 9 January 1902, Folder 245/18E: "Francis Galton to Karl Pearson, 1901-1902," Galton Collection. For the year of Galton's parents' wedding, see "Plate XX," opposite p. 26, text under portrait of "Frances Anne Violetta Darwin" (Galton's mother), Pearson, *Life of Galton*, vol. 1; Forrest, *Francis Galton*, 4. Galton's grandfather had been disowned by the Society of Friends in 1795 "for fabricating and selling instruments of war," but, characteristically, he ignored the resolution and continued attending meetings until his death in 1832. See Pearson, *Life of Galton*, vol. 1, 45. For useful information on this matter and good leads to more, see Leonore Davidoff and Catherine Hall, *Family Fortunes: Men and Women of the English Middle Class, 1780-1850* (Chicago: University of Chicago Press, 1987), 102-03, 484.

As Raymond Fancher has noted, he was "virtually *never* mentioned in family diaries except in the context of his education or intellectual exploits."[9]

Accordingly, the tremendous weight thrust onto Galton's infant shoulders is worth considering: not only was he expected to qualify for university honors, he was also made to understand that he should aspire to equal his illustrious grandfather who was regarded in the family circle and beyond as an intellectual giant. One might think that with so serious a burden assigned at such a tender age, some self-conceptual strain might have been expected. At age four he seems to have regaled his family with the announcement that he was saving his pennies "to buy honours at the University." Moreover he appears to have required frequent reassurance as to the strength of his abilities. Quite early on he began impressing family members with remarkable-seeming feats of memory. At age five, when permitted to assist in the rounding-up of geese, he staggered back clutching a flapping gander by the throat, declaiming, to his mother's delight, the lines from "Chevy Chase" "Thou art the most courageous knight/ "That ever I did see." When he was the same age, a housemaid sent to conduct him home from school came upon him holding off a group of teasing boys, reciting the lines for her to report "Come one, come all, this rock shall fly,/ "From its firm base, as soon as I." When at an unspecified young age he fell from his pony into a muddy ditch, he had at the ready for his brother Darwin the lines from "Hudibras" "I am not now in Fortune's power/ "He that is down can fall no lower."[10] Through such processes and counter-machinations, he seems to have come to conceive of himself as equal to his great life's challenge. Then again, what manner of

---

[9] Pearson, *Life of Galton*, vol. 1, p. 63, 65-70; Forrest, *Francis Galton*, 4-9; Fancher, *Intelligence Men*, 19-22. One of Pearson's sources, Galton's sister Elizabeth, testifies that Adèle taught herself Latin and Greek so as to instruct her brother (*Life of Galton*, vol. 1, p. 63), but Fancher seems correct in believing the acquisition to have consisted of "some Latin and French" (*Intelligence Men*, p. 20). For his characterization of the family diaries, see p. 21.

[10] For the penny-saving anecdote, see Pearson, *Life of Galton*, vol. 1, 69n. For the poetry performances, ibid., 64 for all.

choice did he have?

Eventually, things began to come undone. One obvious misstep came when he cut short his university preparations at sixteen to become an indoor pupil at Birmingham General Hospital, and to study, in the next year, at King's College Medical School in London. (Reportedly, his mother had insisted that his education should duplicate that of her father.) When Galton interrupted his medical apprenticeship at eighteen to pursue a degree in mathematics and his long-anticipated honors at Trinity College, Cambridge, the deficiencies of his preparatory education and the ill-advisedness of his detour into medicine began to make themselves apparent. He took a third class in the May examination, but this poor showing was attributed to a three-month absence due to illness. The second year started out with his showing some slight improvement but also with the first onset of debilitating headaches and dizziness. In March he got a second class in an important examination in which some of his friends who had been tutored identically had taken firsts. At the opening of his third year Galton suffered a severe and disabling emotional breakdown. Eventually he was compelled to abandon his pursuit of mathematics and of honors altogether, and to take an ordinary "poll" degree. He had also failed, in his third year, to win any prizes in a university poetry competition, a contest which he seems to have entered in the hope of emulating his grandfather in that arena. After graduation he returned dispiritedly to his medical studies, but when his father died in 1844, Galton was left, at twenty-two, the means to cast off from medicine and to launch upon the period of travel and dissipation which he later termed his "fallow years."[11] It seems safe to speculate that Francis Galton's exalted self-concept had been shattered at Cambridge, that this had been the cause of his breakdown, and that

---

[11] For the medical studies, see ibid., 90-91, 92, 99-119, and Forrest, *Francis Galton*, 9-16; For his Cambridge period, Pearson, *Life of Galton*, vol. 1, 140-81, Forrest, *Francis Galton*, 19-23, and Fancher, *Intelligence Men*, 23-24. For Galton's return to medical studies, see Pearson, *Life of Galton*, vol. 1, 181-93; for the death of Galton's father and his abandonment of medical studies, ibid., 191-95. For Pearson's evaluation of Galton's breakdown and its possible causes, see ibid., 194-95. For Galton's "fallow years," see ibid., 196-210, and Forrest, *Francis Galton*, 27-37.

he had been incapable of salvaging and reconfiguring the pieces.

Gradually, however, given his great wealth, his family connections, and his native curiosity, things improved. In 1849 as he was outfitting himself for a hunting expedition in South Africa, he learned from his engineer cousin Douglas Galton that the Royal Geographical Society was seeking someone to perform a reconnaissance and mapping of the area. Galton consulted with the Society as to its needs, submitted a detailed plan of the journey he proposed, and between 1850 and 1852 conducted a necessarily much-revised execution of his plan in what is now the nation of Namibia. Subsequently he published a memoir for the Society along with a popular book describing his adventures, was awarded one of the Royal Geographical Society's gold medals for 1854, and was elected to its council.[12] In 1856 he was admitted to the Royal Society

---

[12] For accounts of Galton's African expedition, see Francis Galton, *Tropical South Africa* (London: John Murray, 1853) (the second edition of this same work bore the title *Narrative of an Explorer in Tropical South Africa* (London: Ward, Lock & Co., 1889)); Galton, "Recent Expedition into the Interior of South-Western Africa," *Journal of the Royal Geographical Society* 22 (1852): 140-63; Galton, *Memories*, 121-51; Pearson, *Life of Galton*, vol. 1, 211-42; Forrest, *Francis Galton*, 38-54; and Raymond E. Fancher, "Francis Galton's African Ethnography and Its Role in the Development of His Psychology," *British Journal for the History of Science* 16 (1983): 67-79. Note that Forrest explains that Galton's second-in-command, Charles Andersson, stayed behind after Galton returned home and "achieved the two goals which Galton set for himself but never reached," and cites his respective account, C. J. Andersson, *Lake Ngami* (London: Hurst & Blachett, 1856), which seems to provide correctives to, or at least variant perspectives on, Galton's reports.

Galton had meant to proceed north out of Cape Colony for Lake Ngami, but unrest among the Boers had forced him to sail for Walfisch (now Walvis) Bay in southwest Africa instead, so as to explore the land of the Damaras and Ovampos before, as the plan ran, proceeding east to Lake Ngami. It should be noted that much of this territory had already been explored by Europeans since there were Rhenish missionary stations extending some one-hundred-fifty miles inland along a river leading from the coast. Moreover, contrary to Pearson's and Forrest's accounts, and as no modern-day writers on the subject seem to have recognized, much of Galton's own professedly pioneering route into the Damara territory seems also to have been traversed by European predecessors. See Rev. F. N. Kolbe, "An Account of the Damara Country," *Journal of the Ethnographic Society of London* 3 (1854): 1-3. This report was communicated by the Right Hon. Earl Grey, Principal Secretary of State for the Colonies, and read before the Society on January 15, 1851. Why it should have awaited publication until two years after Galton's first report

itself. Around this time Galton was also elected to the Athenaeum Club, and thus began moving in London's most rarefied intellectual circles. Soon he developed a reputation as an expert in the art of travel and developed an interest in meteorological and anthropological subjects. Thus he seems to have been advancing, through a variety of undertakings and associations, on the self-definition of his formative years, before Cambridge. (Pearson would refer to this period as "The Reawakening.")[13]

In 1859 Galton was radicalized, even more than others of the day, by the appearance of his cousin's monumental *On the Origin of Species*. Galton testified in later years that Darwin's book had unburdened him at a stroke of religious encumbrances under which he had been struggling, but the effect may have been even more fundamentally liberating than this.[14]

By 1863 and 1864, the period in which he was formulating his ideas on eugenics, Galton was engaged in a myriad of undertakings. To judge by his publications alone, he was turning out travel guides, reporting to the British Association on measuring devices, developing European weather charts, writing directions for experienced meteorology observers, discoursing on African and Alpine geography, discovering the anticyclone, and considering the prehistory of animal domestication. Moreover, as its honorary secretary, Galton was attending to the meetings and affairs of the

---

of his expedition seems an interesting question in itself.

[13] Pearson's Chapter 7 in *Life of Galton*, vol. 1, "The Reawakening: Scientific Exploration," pp. 211-42, is largely concerned with the African expedition. His Chapter 8, in vol. 2, concerns Galton's "Transition Studies: Art of Travel, Geography, Climate," pp. 1-69.

[14] See Galton to Darwin, 24 December 1869, in Pearson, *Life of Galton*, vol. 1, Plate 2, between pp. 6 and 7, and ibid., 207-08, for Pearson's discussion of Galton's statement reading ". . .your 'Origin of Species' formed a real crisis in my life; your book drove away the constraint of an old superstition as if it had been a nightmare and was the first to give me freedom of thought." See also Galton to Alphonse de Candolle, 5 June 1882: "I can truly say for my part that I was groaning under the intellectual burden of the old teleology, that my intellect rebelled against it, but that I saw no way out of it till Darwin's 'Origin of Species' emancipated me." Pearson, *Life of Galton*, vol. 2, 206. See also Galton, *Memories*, 287.

FIGURE 1. Francis Galton at age 38, in 1860. Reprinted from Karl Pearson, *The Life, Letters, and Labours of Francis Galton*, vol. 2 (Cambridge: Cambridge University Press, 1924), Plate VIII, opposite p. 40.

Ethnological Society of London, a group increasingly concerned with discussing possible causes of the differences observed between the races. He was manifesting, in the upper reaches of the Royal Geographical Society, so persistently disputatious a tendency that he was asked in 1863 to resign an honorary secretaryship, and narrowly missed being ordered by the general membership to vacate his council seat and play no further part in the administration of Society affairs. From 1863 to 1866 he was engaged in publishing, along with Herbert Spencer, Norman Lockyer, Thomas Huxley, John Tyndall, and ten other luminaries a weekly journal of literary and scientific affairs, *The Reader*, which was steadily being torn asunder by the headstrong eminences involved.[15]

It was in the midst of these activities, in 1865, that Galton both invented and introduced eugenics (which he would put off christening and officially unveiling until 1883). Why had he invented such an idea, along with its central claim that intelligence and ability are not developed but inherited? The foregoing account of his life suggests various personal incentives. Perhaps his cousin's epochal achievement had resolved vital questions of self-definition for Galton. Perhaps Darwin's startling ascension into monumental significance had served to persuade Galton that their shared grandfather's genius had indeed been transmissible, that he might

---

[15]For a listing of these (and almost all of Galton's) publications, see Forrest, *Francis Galton*, "Appendix III [:] Bibliography of Galton's published work," 303-17. See also the listing "Memoirs, Papers and Letters to Journals of Francis Galton" which appears in Pearson, *Life of Galton*, vol. 3B, in the index for all four volumes, at pp. 655-68; and the listing in Blacker, *Eugenics*, 329-41, which is based on Pearson's and gives better citations for the entries there. For a discussion of Galton's problems at the Royal Geographical Society, see Forrest, *Francis Galton*, 68-72. Accounts of Galton's experience with the *Reader* appear in Galton's *Memories*, 167-68; Pearson, *Life of Galton*, vol. 2, 67-69; and Forrest, *Francis Galton*, 77. *The Reader*, which existed from 3 January 1863 to 12 January 1867, was an impressively ambitious undertaking even by Victorian standards. It took as its province all of British and much of world literature, science and art; it published reviews of publications ranging from textbooks printed in England to "comic" books; and it affords very useful accounts of meetings of both the Ethnological Society of London and the Anthropological Society. Moreover, it features numerous articles by Galton which appear in none of the published bibliographies of his work.

have the stuff of genius in himself as well, and that his childhood precocity had not been, after all, merely the product of an overly ambitious nurturing. If he could prove that intelligence and ability are inherited, he might fully reclaim, and re-inhabit, his founding and fundamental sense of himself as embodying hereditary genius. Certainly he could have felt personal encouragement as well for his contention that a seemingly superior population breeding carefully within itself could develop an increasing superiority in selected attributes over the general population. Such a process could well have been exemplified for Galton by the industrious and intelligent Quaker ancestors on his father's side. Another such example may have been afforded him by the "intellectual aristocracy" described in our time by Noel Annan.[16] This constellation of intermarrying

---

[16] For a testimonial regarding the influence of the *Origin* on Galton's interest in heredity and eugenics, see Galton, *Memories*, 288.

Note that Pearson attributed various of Galton's qualities to his Quaker heritage. See [Pearson,] "Francis Galton" (obituary), 441, where he explained that "not only did the Society of Friends unite men religiously but it produced special temperaments genetically. . . ..almost a biological type." (It might be noted that Pearson was himself descended from Quaker forebears, although he certainly lacked various of the attributes which he was assigning to their genetic influence.)

Annan, "The Intellectual Aristocracy," in ed. J. H. Plumb, *Studies in Social History* [:] *A Tribute to G. M. Trevelyan* (London: Longmans, Green and Co., 1955), 241-87. Note that Annan's study much resembles, on a structural level, Galton's in *Hereditary Genius* and the two articles building to it insofar as it treats the transmission of ability in similar, and sometimes the same, families. But Annan, quite unlike Galton, stresses the influence of family environment, example, and tradition:

> As infants they had learnt by listening to their parents to extend their vocabulary and talk in grammatical sentences. . . .When older they subconsciously apprehended from hearing discussions between their elders how to reason logically. They lived in houses in which books were part of existence and the intellect was prized. . . .The successful children gained by acquiring the habit of thinking accurately at an early age. (250-51) Clearly certain families produce a disproportionately large number of eminent men and women. But equally clearly this study shows that men of natural but not outstanding ability can reach the front ranks of science and scholarship and the foremost positions in the cultural hierarchy of the country if they have been bred to a tradition of intellectual achievement and have been taught to turn their environment to account. Schools and universities can so train young men, but such a training has a far stronger command over

middle-class families, which had been waxing more illustrious by the generation, included the Galtons, the Darwins, and the Butlers, whom Galton allied himself with in 1853, when he married Louisa Butler, a daughter of a famous headmaster of Harrow and Dean of Peterborough.

Other personal-cum-social influences on Galton's decision to devise a program for artificial selection may include the idea's general currency in the charged air following the thunderclap sounded by the *Origin*. Certainly he was well acquainted with the prominent sanitarian William Farr, who circa 1860, was calling both for surveys of the national intelligence and for programs for the artificial selection of man.[17] Galton may simply have wanted to be the first, among several interested parties, to introduce such a system. Or conceivably he could have felt an intellectual compulsion to do so, borne along on a flood of enthusiasm for the correctness of natural selection, which his 1865 article on animal domestication shows him to have well appreciated and generally understood. Similarly, he showed himself, in the opening pages of

the personality when it is transmitted through a family tradition. (284-85).

[17]Note in this regard Julian Huxley's observation that "With Darwin and the acceptance of the theory of Evolution, the human race realised, in a way impossible to it before, that man was in control of his own destinies, and could effect alterations in his own nature." See Huxley, "The Case for Eugenics," *Sociological Review* 18 (October 1926): 283.

For useful discussions of Farr, see John M. Eyler, *Victorian Social Medicine: The Ideas and Methods of William Farr* (Baltimore: Johns Hopkins University Press, 1979); Victor L. Hilts, "William Farr (1807-1883) and the 'Human Unit,'" *Victorian Studies* 14 (December 1970): 143-50; Hilts, *Statist and Statistician* (New York: Arno Press, 1981; a rebinding of his doctoral dissertation in History of Science, Harvard University, 1967, which bore the subtitle *Three Studies in the History of Nineteenth Century English Statistical Thought*), 218-39. Lyndsay Farrall, *The Origins and Growth of the English Eugenics Movement, 1865-1925* (New York: Garland, 1985; a rebinding of his 1969 Indiana University doctoral dissertation), 10-33, discusses a number of other English social analysts and scientists who were considering the importance of natural selection for man in this general period. Finally, see also the catalogue of American popular appeals for directed evolution provided by John Humphrey Noyes in his *Essay on Scientific Propagation* (Oneida, NY: published by Oneida Community, 1870?), 3-5.

"Hereditary Talent and Character" and *Hereditary Genius*, to have
been keenly alert to the successes enjoyed by professional breeders
in demonstrating the genetic plasticity and ductility of animal
species.[18] Finally it should be noted that a broad variety of other
psychohistorically-emanating and socially-directed impulses have
been suggested by modern writers as well. These have ranged from
his "lingering self-doubt" regarding "the validity of his own
success,"[19] and the impulse to create a substitute for the religion he
had given up,[20] or to compensate for his own inability to produce
offspring,[21] to an apparently irresistible need to express "his
independence of spirit, his developed critical powers, his love of
measuring and assessing things, [and] his essential trustfulness and
benevolence," as one of his more filiopietistic biographers has
conjectured.[22]

---

[18] "The First Steps towards the Domestication of Animals," *Transactions of the Ethnological Society of London* 3 (1865): 122-38; "Hereditary Talent and Character," "Part I," 158, 165; *Hereditary Genius*, 1.

[19] Daniel J. Kevles, "Annals of Eugenics[:]A Secular Faith—I," *New Yorker*, 8 October 1984, 56. This article was the first installment of a four-part series which was continued in the numbers for 15, 22, and 29 October 1984.

[20] Forrest, *Francis Galton*, 85; Ruth Schwartz Cowan, "Nature and Nurture," 158-64.

[21] Forrest, *Francis Galton*, 85; Daniel J. Kevles, *In the Name of Eugenics: Genetics and the Uses of Human Heredity* (Berkeley: University of California Press, 1985), 9; Kevles, "Annals," "Part I," 56, where he speculates that "Galton may well have diverted frustration over his own lack of children into an obsession with the eugenic propagation of Galtonlike offspring."

[22] Blacker, *Eugenics*, 104. Cowan expresses a similar view in "Nature and Nurture," p. 153, observing that eugenics "provided him with a philanthropic cause that could engage all his altruistic and scientific impulses." She also goes on to assign to Galton general political purposes of a reformist cast, stating that "the proposals for social reform that Galton first enunciated in 'Hereditary Talent and Character' fulfilled his need to play the role of a social philanthropist" (157).

CHAPTER FOUR
THE POLITICAL MATRIX

ALL THIS MODERN biographical consideration certainly serves to honor George Sarton's dictum that when treating "a man's discovery, one must explain. . .why it was he who made it. . . ."[1] We have grappled mightily with the question of why *Francis Galton* invented eugenics, of what in his life experience might have impelled him to articulate a system of artificial selection. What modern inquiry has not addressed for the most part is why Francis Galton invented *eugenics* — why he molded artificial selection into the carefully constrained form he made it take.

This second sort of question Galton's American legatees were greatly more disposed, and somewhat better equipped, to ask. Obviously they shared his times and understood them far more intimately than we can. They shared many of his scientific concerns, as fellow differential psychologists, and quite possibly his political ones as well. Spared the distraction which might have been posed by overmuch information on his life history, they were all the more inclined to connect salient features of his system to the political topography that stretches beneath all social development. Moreover they seem to have felt the necessity of such inquiry, and in this they were not alone. Karl Pearson, Galton's anointed successor, would advise his students that ". . . it is impossible to understand a man's work unless you understand something of. . .the state of affairs social and political of his own age." "You might think it possible to write a history of science in the 19th century and not touch theology or politics," he continued; "I gravely doubt whether you would come down to its actual foundations. . . ."[2]

---

[1]Quoted as one of several epigraphs at the opening of James W. Tankard, Jr., *The Statistical Pioneers* (Cambridge, MA: Schenkman Publishers Inc., 1984), iv.

[2]Pearson, *The History of Statistics in the 17th and 18th Centuries Against the Changing Background of Intellectual, Scientific and Religious Thought: Lectures by Karl Pearson Given at University College London during the Academic Sessions 1921-1933*, ed. E. S. Pearson (London: Charles Griffin & Co., 1978), 360. The author was directed to this statement by Theodore M. Porter, *The Rise of Statistical Thinking, 1820-1900* (Princeton: Princeton University Press, 1986), 9.

For other analysts perceiving political factors as conditioning Galton's development of eugenics, cf. Ruth Schwartz Cowan, "Sir Francis Galton," dissertation, 55-62 especially; Cowan, "Sir Francis Galton and the Continuity of

Attempting to comprehend the American hereditarian psychol-ogists' general view of Galton requires that some historical scaffolding be erected so as to attain their approximate attitude (in several senses of the word) toward him and thus effect the *tour d'horizon* which they seem to have taken of him and his system of eugenics generally. Of necessity this skyline must itself be heuristically composed inasmuch as we cannot know with assurance which of the uplifted features to be considered were the ones which they as individuals made out. Fortunately we can surmise with some confidence the general orienting sense of Galton's motives shared by many American hereditarians, and be assisted to our positioning thereby. Access to this information is provided by the American comparative psychologist Robert Yerkes, who, five years before his influence reached its highest point, exposed a brilliantly revelatory perspective of Galton.

In 1912, when speaking before the National Conference of Charities and Correction, and in subsequent deliveries of what seems to have been the same speech nationwide, Yerkes expressed the opinion that in creating eugenics "Sir Francis Galton. . . . was probably to no greater extent influenced by Darwin than by Plato." What could Yerkes have meant by this? He had opened his address with a reference to Plato's eugenic utopia the *Republic*, observing that "during the centuries since, many able individuals have thought with him, have accepted his ideals, and have longed for certain of the social results of the eugenics measures. . . ."[3] But which ideals

---

Germ-Plasm: A Biological Idea with Political Roots," *XIIe Congrès International D'Histoire des Sciences (Paris 1968)*. Actes Tome VIII, Histoire des Sciences Naturelle et de la Biologie (Paris: Albert Blanchard, 1971), 181-86; Cowan, "Nature and Nurture," 154-57; and Allan R. Buss, "Galton and the Birth of Differential Psychology and Eugenics: Social, Political, and Economic Forces," *Journal of the History of the Behavioral Sciences* 13 (January 1976): 47-58.

[3] Yerkes, "Eugenics: Its Scientific Basis and Its Programme," "An Address before the Eugenics Section of the National Conference of Charities and Correction," given on 19 June 1912 at the 39th annual meeting of the Conference, held at Cleveland, Ohio, 12-19 June 1912. Only an abstract of this address appears in the *Proceedings of the National Conference of Charities and Correction* 39 (1912): 279-80. Yerkes's manuscripts for the speech survive in several formats, all of which suggest that the first reference to Plato came at pp. 1-2 and the reference to Plato's influence on Galton on p. 3. See Robert M. Yerkes Papers, Folder 1129, Box 59, Series II, Record

and what social results? Surely Yerkes was not referring, for his own part, to such social features as the abolition of the family that Plato's scheme had prescribed for the ruling class, but, instead, it would seem, to the tripartite hierarchy which such a measure was meant to facilitate. It was widely perceived that Plato's tract had been composed in large part as a response to the Athenian democratic polity which his oligarchic family and class had briefly overthrown, and which he instrumentally despised.[4] His *Republic* presented a system in which all policy is dictated by a ruling class of experts, the Guardians, who suffer no scrutiny or interference from the governed classes beneath them, the enforcing Auxiliaries and the broad working class of farmers and craftsmen at the bottom. Thus Plato's *Republic*, like many of his strictures therein and elsewhere, was devoted to precluding change, to freezing the social wheel with the better sort on top.

To judge by the American hereditarians' own experience, it seems probable that the feature of Plato's utopia which struck them most forcefully, and which some saw strikingly reflected in Galton's

---

Group 569, Manuscripts and Archives, Sterling Memorial Library, Yale University, New Haven, Connecticut.

[4] Obviously this characterization of Plato's utopia accords with that of Sir Karl R. Popper, *The Open Society and Its Enemies*, 2nd (revised) ed., (London: Routledge and Kegan Paul, 1952) vol. 1, 45-54, 86-113, 119, 120-68. Yet such a view was extant near the opening of the twentieth century as well. E.g., see Sir Ernest Barker, *The Political Thought of Plato and Aristotle* (New York: G. P. Putnam's Sons, 1906), 61ff., 83-162. Barker explained that "ignorance was to Plato the especial curse of democracy. Here, instead of the professional, the amateur and the sciolist were predominant. In Athens especially, democracy seemed only to mean the right divine of the ignorant to govern wrong" (87-88). Indeed, "indignation makes the *Republic*" (85). See also Barker, *Greek Political Theory: Plato and His Predecessors* (London: Methuen & Co., 1918), 145-270. Even so pronounced an admirer of Plato as the Oxford tutor Richard Lewis Nettleship conceded that for all of Plato's assurances to the contrary, his "system. . .would apparently not admit of promotion from the lower class. . . ." See Nettleship, *Lectures on the Republic of Plato*, 2d ed., ed. Lord Charnwood (London: Macmillan and Co., 1901), 135n.

It should be noted, too, that Barker, and perhaps some others of the time, did not regard Plato's production as having been written as a utopia, "a city in the clouds, a sunset fabric seen for an hour at evening and then fading into night": "It is based on actual conditions; it is meant to mould, or at any rate to influence, actual life." See Barker, *Greek Political Theory*, 239.

eugenics, was Plato's mechanism for justifying his system—the "necessary fiction" which has come to be known as the Noble Lie. Socrates reveals this strategy of deception to Glaucon (one of Plato's older brothers) only after exhibiting considerable embarrassment as to its nature (". . . I hardly know where I shall find the courage or where the words to express myself," he exclaims):

> We shall tell our people, in mythical language: you are doubtless all brethren, as many as inhabit the city, but the God who created you mixed gold in the composition of such of you as are qualified to rule, which gives them the highest value; while in the auxiliaries he made silver an ingredient, assigning iron and copper to the cultivators of the soil and the other workmen.[5]

Shortly thereafter Plato has his spokesman propose an open meritocracy in which all of superior quality would be discovered and admitted to the superior classes. But inasmuch as he provides elsewhere for the children of the Guardians to be secretly set apart at birth, so that "the breed of the Guardians. . .be kept pure," and to be specially educated toward Guardianship from the first, such an assurance rings hollow.[6]

How might all of this be perceived as significant to Galton's development of eugenics? First, there was one obvious and massive parallel shared by Plato's Athens and Galton's England. To understand this similarity, an extensive description of the political background to the creation of eugenics is helpful.

The overwhelming domestic political concern for England in the 1860s and for some decades beyond was the seemingly imminent extension of the franchise to the lower classes. This was an idea that had suddenly—and altogether unexpectedly—metamorphosed from a debatable likelihood into an onrushing inevitability. As Thomas Carlyle observed, "the wisest prophecy finds it was quite wrong as

---

[5] Plato, *The Republic*, III.414-15. The edition employed is *The Republic of Plato Translated into English with an Analysis and Notes*, trans. John Llewelyn Davis and David James Vaughan (London: Macmillan and Co., 1891), 113-14.

[6] V.460; ibid., 168.

to date; and. . .is astonished to see itself fulfilled, not in centuries as anticipated, but in decades and years. . . ."[7] There had been two precipitants for this. The first was the increasingly undeniable indication, from the Great Exhibition in 1851 on, that England was falling behind Germany and the United States in industrial competitiveness, and particularly in the capacity for effecting technological innovation.[8] Every successive international

[7]Carlyle, "Shooting Niagara," *Macmillan's Magazine*, August 1867, 320. This, Carlyle's last essay, was soon reprinted as a pamphlet with the title *Shooting Niagara and After* (London: Chapman and Hall, 1867).

[8]The Great Exhibition, for which the Crystal Palace was erected in Hyde Park, was the idea of Prince Albert and in this he was greatly assisted by the engineer John Scott Russell and the chemist and eventual M. P., Lyon Playfair, First Baron Playfair of St. Andrews. Albert had planned the Exhibition—in 1849 with Russell—as the world's first international display of manufactures, but more than this, his purpose seems to have been to demonstrate to complacent Britons the degree to which they were falling behind in technological innovation; apparently an associated concern for Albert was generating public support for both a much-improved system of education for working-class children and a program of industrial education. See Theodore Martin, *The Life of His Royal Highness The Prince Consort*, 2d. ed., 5 vols. (London: Smith, Elder & Co., 1875-79), vol. 2, 224-25, for information on the collaboration of Albert, Russell and three others. See also [Albert Samuel Hewins,] "Russell, John Scott," *Dictionary of National Biography* 17, Robinson-Shares, eds. Sir Leslie Stephen and Sir Sidney Lee (London: Oxford University Press, 1917), 465-66; and [Arthur Harden,] "Playfair, Lyon," *Dictionary of National Biography Supplement* 22, eds. Sir Leslie Stephen and Sir Sidney Lee (London: Oxford University Press, 1917), 1142-44. See Albert, Prince Consort, *Letters of the Prince Consort 1831-1861*, ed. Kurt Jagow (London: John Murray, 1938) for Albert to Col. Sir Charles Phipps, 20 September 1849, for his spelling out of his educational concerns, which he saw as the first of four steps toward the "improvement of [the] conditions" of "the working classes (so called)" (154). It is clear that Albert saw many of England's present and future problems regarding international industrial competitiveness as deriving from the exceptionally pronounced divisiveness and exclusivity that characterized English society. See in this regard Reginald Pound, *Albert: A Biography of the Prince Consort* (London: Michael Joseph, 1973), especially p. 288, where Pound cites Lady Augusta Bruce's recollection of Albert's remarks to her in 1855 on the disastrous effects such a mindset was having on the British diplomatic service, which, she says, he described as "'the worst.'"

Naturally there were apprehensions antecedent to the Great Exhibition and to Albert's recognitions as well. See Joseph Kay (later Kay-Shuttleworth), *The Social Condition and Education of the People in England and Europe; Shewing the Results of the Primary Schools, and of the Division of Landed Property, in Foreign Countries*, 2 vols. (London: Longman, Brown, Green and Longmans, 1850). Asa Briggs quotes the

Economist as having warned "even before the Great Exhibition," that " 'the superiority of the United States to England is ultimately as certain as the next eclipse.' " See Briggs, *The Age of Improvement, 1783-1867* (New York: David McKay Co., 1959), 400. Moreover, W. H. G. Armytage quotes Richard Cobden as prophesying as early as 1835 that

> it is . . .from the silent and peaceful rivalry of American commerce, the growth of its manufactures, its rapid progress in internal improvements, the superior education of its people, and their economic and pacific government. . .that the grandeur of our commercial and national prosperity is [most] endangered.

As Armytage notes, Cobden was only "an unknown thirty year old Manchester calico printer" at the time. See Armytage, "W. E. Forster and the Liberal Reformers," in ed. A. V. Judges, *Pioneers of English Education: A Course of Lectures Given at King's College, London* (London: Faber and Faber, 1952), 207.

The effects of the Great Exhibition were both dramatic and immediate. Apparently the American inventions on display that caused the deepest stir were the sewing machine, the revolver, and the mechanical reaper, all assembled with interchangeable parts. See George Haines IV, "Technology and Liberal Education," in eds. Applemann et. al., *1859*, 102. The degree to which Prince Albert's plan for arousing the English people was successful is indicated by a contemporaneous account of the Exhibition's general effect:

> No amount of oral evidence or of official documents or of statistical facts, however for years reiterated and republished, would have produced the same strength of conviction in the public mind which a few weeks' display in the world's bazaar, in Hyde Park, effected. The nation at once saw and admitted its partial failure.

This appears to be the testimony of P. Le Neve Foster, Secretary of the Society for the Encouragement of Arts, Manufactures, and Commerce; see *Addresses Delivered on Different Public Occasions by His Royal Highness The Prince Albert, President of the Society for the Encouragement of Arts, Manufactures, and Commerce* (London: Bell and Daldy, 1857), 90-91. For an attempt to give a less alarming aspect to the results of the Exhibition, see Henry Cole, "On the International Results of the Exhibition of 1851," in *Lectures on the Results of the Great Exhibition of 1851, Delivered Before the Royal Society of Arts, Manufactures, and Commerce At the Suggestion of H.R.H. Prince Albert, President of the Society*, 2d series (London: David Bogne, 1853), 419-51. In the end, Cole was reduced to remarking, ". . .I look with no alarm at the progress of science abroad. . ., if our neighbors produce science, and we want it, we shall be able to obtain it from them on equitable terms, and turn it to account." "After all," he ventured, ". . .if we supply the practical execution, and our neighbors the philosophical theory, it may. . .be only a proper division of labour between friends." See also, for a well-reasoned modern consideration of the social-psychological effects on the English of the Great Exhibition, Donald K. Jones, *The Making of the Education System, 1851-81* (London: Routledge & Kegan Paul, 1977),

competition and inspection tour heaped the fearful news higher. As early as the Paris Exhibition of 1855, Lord Ashburton warned that, should things continue as they were, "this England of which you are so justly proud, this storehouse of nations, 'Oh it will be a wilderness again, peopled with worse than wolves.' Peopled with starving desperate outlaws."[9] The second precipitant was the

---

2-4. See also in this regard, P.W. Musgrave, *Technical Change: The Labour Force and Education; A Study of the British and German Iron and Steel Industries 1860-1964* (Oxford: Pergamon Press, 1967), 52-53. Finally, for an understandably celebratory American discussion of the competition, see Charles T. Rodgers, *American Superiority at the World's Fair* (Philadelphia: John J. Hawkins, 1852).

[9] After the even more disturbing international exhibition held in Paris in 1867 the Conservative government appointed a select committee to investigate matters and allay fears. As W. H. G. Armytage notes, this committee discovered that Belgian girders were being used throughout Britain, and that not only centrifugal pumps, but also screw lathes at British factories, and the machinery-making weapons at the Birmingham Small Arms Factory (formerly owned by the Galtons), even the printing presses at the London *Times* and the leading Manchester newspapers which printed this disturbing news, had all come from America. See Armytage, "W. E. Forster," 208-09. See also one of the most remarkable English books of the nineteenth century, J. Scott Russell, *Systematic Technical Education for the English People* (London: Bradbury, Evans, & Co., 1869) for a discussion of the lessons and effects of the Great Exhibition of 1851, the Great Exhibition held, in Paris, in 1855, the "humiliating" English Exhibition of 1862, and the Exhibition of 1867, in Paris, where "we were beaten—not on some points, but by some nation or other on nearly all those points on which we had prided ourselves." Russell's book, which abounds with the verbatim testimony of a wide range of experts, also recounts his own experiences in the course of a tour of Prussia, which nation he describes as "an organised democracy, and. . .not, as we imagined governed by the King and an aristocracy." With its enthusiastic description of the process and the benefits of Prussia's eighteen-year conversion from aristocracy to democracy—largely effected through the education of all levels of the population—and with Russell's effusions on the joys of interacting with an intelligent and broadly educated Prussian citizenry at all levels, *Systematic Technical Education* represents nearly an exact antithesis to *Hereditary Genius*, which, of course, appeared in the same year. See also J. Scott Russell, "Technical Education a National Want," *Macmillan's Magazine*, April 1868, 447-59, a forerunner to the book. See also, for a reprinting of a letter from Lyon Playfair printed in the London *Times* regarding the 1867 Exhibition and for two defenses of England's status also printed there in response, "The Paris Exhibition and Industrial Education," *Journal of the Society of Arts*, 6 July 1867, 477-79.

For Lord Ashburton's remarks, which were made in the course of an address

recognition that this problem dictated a vastly improved technical education for the English industrial workforce and, beneath this, as an enormous prerequisite, the establishment of a nationwide system of universal and compulsory elementary education. To remain in the great race of nations, England must finally educate what Richard Cobden had described in 1851 as "the most ignorant Protestant people on the face of the earth." As the *London Review* observed in 1861, the idea of educating the working classes "to enable them to occupy. . .the sphere of manual labor intelligently. . .is now admitted on all hands to be a crying necessity."[10]

given "At the Banquet in the Birmingham Town Hall, On the Occasion of Laying the First Stone of the Birmingham and Midland Institute," on 22 November 1855, see *Addresses Delivered by Prince Albert*, 173.

[10]Cobden's statement was quoted by Armytage, "W. E. Forster," 207. "Report of the Commissioners Appointed to Inquire into the State of Popular Education in England. London 1861," *London Review*, July 1861, 504.

For a helpful contemporaneous discussion of the evolution of public attitudes on public education from the end of the eighteenth century, see "Popular Education: The Means of Obtaining It," *The Prospective Review: A Quarterly Journal of Theology and Literature* 8 (1852): 16-56, where it is observed that "a mere list of . . .the numerous pamphlets which have been poured forth from the press during the last few years on the topic of Popular Education. . .would form a not inconsiderable catalogue," and that "if there be any who. . .are still opposed to the doctrine [of universal education], their voices are now seldom heard, and never above a whisper." For a modern consideration of why universal and compulsory education was not established in the years 1840-1860, see Geoffrey Best, *Mid-Victorian Britain, 1851-1875* (New York: Schocken Books, 1972), 156-57. In this regard, see also Jones, *Making of the Education System*, passim.

For invaluable contemporaneous accounts – and examples – of the evolving support for universal and compulsory or industrial education in the 1860s, see the following: "Report of the Commissioners Appointed to Inquire into the State of Popular Education in England. Vol I. Presented to Parliament in April, 1861," *British Quarterly Review* 34 (July 1861): 218-33, an article which advocated "infant schools" for children of working classes much like modern "headstart" programs. "Art. I." *Edinburgh Review*, July 1861, 1-20, on the report of a commission touring European educational institutions after the Paris Exhibition of 1855. "Popular Education in Prussia," *Westminster and Foreign Quarterly Review* 21 (1 January 1862): 169-200. "Endowed Schools," *Westminster and Foreign Quarterly Review* 21 (1 April 1862): 340-57. John Morley, "Social Responsibilities," *Macmillan's Magazine*, September 1866, 377-86. Thomas Huxley, "A Liberal Education; And Where to Find It. An Inaugural Address," *Macmillan's Magazine*, March 1868, 367-78, where

Thus the rub: Once educated, this cohort would undoubtedly become more insistent on its right to political participation. The overriding problem here was one of simple numbers, with the lower classes constituting more than three-quarters of the population.[11] No

he questioned "if it be wise to tell people you will do for them out of fear of their power, what you have left undone, so long as your only motive was compassion for their weakness and their sorrows" (367-68). "Popular Education," *Westminster and Foreign Quarterly Review* n.s. 33 (1 April 1868): 421-41, which affords a useful characterization of attitudes held by those of the "ruling classes" who had been resisting educational reform. Dudley Campbell, "Compulsory Primary Education," *Fortnightly Review*, 1 May 1868, 570-81. "Primary Education," *Westminster Review* n.s. 35 (1 April 1869): 458-83, which discusses all the work to be done if a viable system were to be instituted. Dudley Campbell, "Compulsory Education," *Westminster Review* n.s. 36 (1 October 1869): 550-56. Rev. Canon Norris, *The Education of the People* (London: Macmillan and Co., 1869), for a well-noted statement by an opponent of compulsory education.

For useful modern discussions of various legislative and investigative documents, see Celina Fox, Richard Johnson, Roy MacLeod, Edward Miller, and Gillian Sutherland, *Education*, ed. Gillian Sutherland (Dublin: Irish University Press, 1977). See Musgrave, *Technical Change*, 52-55, for a good review of investigations and legislation regarding education reform for 1851-1870. For the subject of technical education, see Michael D. Stephens and Gordon W. Roderick, "British Artisan Scientific and Technical Education in the Early Nineteenth Century," *Annals of Science* 29 (June 1972): 87-98; Stephens and Roderick, "Science, the Working Classes and Mechanics' Institutes," *Annals of Science* 29 (December 1972): 349-60; Stephens and Roderick, "Changing Attitudes to Education in England & Wales 1833-1902: The Governmental Reports, with Particular Reference to Science & Technical Studies," *Annals of Science* 30 (June 1973): 149-64; Stephens and Roderick, "American and English Attitudes to Scientific Education during the Nineteenth Century," *Annals of Science* 30 (December 1973): 435-56.

With respect to the actual record of progress in instituting universal and compulsory education in England, see Haines, "Technology and Liberal Education," who notes that:

> Though as early as the 1850's the need for a compulsory and national system of education was urged, not until A. J. Mundella's Act in 1880 was the former achieved. As for a national system, even of the Act of 1902 a recent English historian of education observes: 'Although it would be an exaggeration to claim that the Act created a national system of education, yet one can justifiably assert that it laid the foundation on which a system could be built' (105).

[11]See R. Dudley Baxter, *National Income-The United Kingdom* (London: Macmillan and Co., 1868). On pp. 15-16, Baxter provides statistics indicating that the "manual

group was more alarmed by the new reality than the middle classes. Though the issue might have been seen as one of simple equity, given that group's own dramatic gains in social mobility and their imminent victories in opening the universities and the civil service to all qualified persons, such was hardly the case. Many in the middle classes saw equity for the lower classes as a threat to the survival of their own. After all, the middle classes of 1865, Galton's cohort, had reason to think themselves significantly threatened on two fronts. Looking upward they could see their own ceiling, the membrane separating themselves from the upper classes, thickening at alarming speed. This also unexpected phenomenon was a result of the enormous growth, both in scale and complexity, of British economic and industrial enterprise. Because it had become increasingly difficult or unappealing for magnates to extricate themselves from the firms they had established, in the words of Harold Perkin "the old cycle of business life, by which the successful moved out to the land, to be replaced by newcomers from below, was in an increasing number of cases broken, a development which helps to explain the declining social mobility of the Victorian age."[12]

---

labour class" amounted to 76.81 percent of the total population of England, 78.05 percent of that of Scotland, 81 percent of that of Ireland, and 77.72 percent of that of the United Kingdom generally, all in 1867. On p. 8 he provides statistics indicating that with respect to "Persons with Income or Wages. England and Wales, 1861," the working classes contributed 79.08 percent of the total. Given that the dependent classes might be expected to be denied the franchise for some time, presumably the last-mentioned figure best represents what proportion the lower classes would be expected to represent of the electorate if given the vote.

[12]For information on the movement for civil service reform, see Briggs, *Age of Improvement*, 443-44. Briggs also cites John Ruskin's observation in his *Pre-Raphaelitism* that for the middle classes it had become a matter of shame not to rise in society; see ibid., 404. For a surefooted discussion of the middle classes' opening of medical education to non-Church-of-England aspirants, one which emphasizes the interpenetrations of developmental theory and ideology, see Adrian Desmond, "Lamarckism and Democracy: Corporations and Corruption and Comparative Anatomy in the 1830s," in ed. James R. Moore, *History, Humanity and Evolution: Essays for John C. Greene* (Cambridge: Cambridge University Press 1989), 99-130. Much of the middle classes' anxiety about the working classes seems indicated by the content, and the popularity, of novels treating that element. See, for example, Benjamin Disraeli, *Sybil, or the Two Nations*, 3 vols. (London: H. Colburn, 1845);

And, as if training their ears on the flooring beneath them, the middle classes could hear all-too-familiar sounds rising. The middle classes' greatest incentive to fear was that of their own example. What is more, their apprehension addressed several possible outcomes: either the newly educated children of the working classes might climb upward, displacing the scions of the current occupants of the professions and other seats of power, or the huge working classes might take over the country, politically, altogether. The middle classes might lose either their patrimony or their hegemony, or both concomitantly.

It should be further stressed that a majority-rule democracy was the one thing that very few above the working classes wanted. Frederic Harrison, the great positivist spokesman and representative liberal of his time, usefully indicated the breadth of the fear which prevailed that England's political system would lurch toward that of America. As Harrison explained in 1873, Conservatives and Liberals alike shared "the profound conviction. . .that whatever their professions might require, in practice it would be madness to abandon Government to the hands of numbers."[13] Harrison's

---

Elizabeth Gaskell, *Mary Barton*, 2 vols. (London: Chapman and Hall, 1848); Rev. Charles Kingsley, *Alton Locke, Tailor and Poet: An Autobiography* (London: Chapman and Hall, 1850); George Eliot, *Felix Holt, The Radical* (Edinburgh and London: W. Blackwood and Sons, 1866). For a thoughtful discussion of these novels and tendencies, see John Gerhard Hessler, "Victorians and the Threat of Democracy," Ph.D. diss., Stanford University, 1977. For an impressive study of who was reading these novels, consult Darko Suvin, "The Social Addressees of Victorian Fiction: A Preliminary Inquiry," *Literature and History* 8 (Spring 1982): 11-40.

Harold Perkin, *The Origins of Modern English Society: 1780-1880* (London: Routledge & Kegan Paul, 1969), 431.

[13]Harrison, "The Revival of Authority," *Fortnightly Review*, 1 January 1873, 5. "The political problem of our age," Harrison also remarked, ". . .is [to] found Authority without oppression upon a Public Opinion without Democracy" (2). Harrison also declared that

> From the opening of 1848 to the present day, the same thing has been shown in fifty movements and on fifty battlefields. . . .This generation is slowly learning that if the social revolution (which grows more and more inevitable) is to be worked out naturally and healthfully, it will never be by Democracy as its method [14].

characterization of a Liberal aversion to universal franchise was exemplified in Walter Bagehot's efforts to point out the absurdity of democracy in his 1859 study of parliamentary reform, and by J. S. Mill's *Thoughts on Parliamentary Reform*, published in the same year. In the latter treatise, the quintessential liberal assured that "no lover of improvement can desire that the *predominant* power should be turned over to persons in the mental and moral condition of the English working classes." That conservatives shared this sentiment need not be demonstrated.[14]

Moreover, many public figures were intent on doing more than merely bemoaning the inevitability of turning their nation over to the masses, whom many referred to, with as much foreboding as sarcasm, as "our future masters."[15] More than two hundred eminences from the fields of religion, education, government, literature and science affixed their names to a memorial to Lord Palmerston published in the *Times* in 1857, calling for the institution of an "educational franchise." This was to be a device for conferring on all "who have the advantage of a liberal education. . .a distinct and separate representation" so that "their votes may not be swamped" by "large popular constituencies."[16] In the same year

Another of Harrison's observations may be pertinent to a general consideration of Galton's possible motives: "[The Tories] dare not, in common decency, put forward the principle which is doubtless their secret faith. . .that the incapable should leave Government to the capable. . . .A governing class ought, at least, to believe that they are specially fit to govern, as our civil and military staff in India do, and as even the Prussian Junker does. But the notion that nature had specially designed them to govern would be too whimsical to be seriously accepted by our conservatives at home. . . ." (5-6).

[14]Bagehot, *Parliamentary Reform-an Essay Reprinted, with Considerable Additions, from the National Review* (London, 1859).

Mill, *Thoughts on Parliamentary Reform* (London: Parker and Son, 1859), 30. This statement appears on p. 327 of Mill, *Essays on Politics and Society*, vol. 2, ed. J. M. Robson (Toronto: University of Toronto Press, 1977).

[15]Briggs attributes this term to Robert Lowe, an M. P. whom Armytage depicts as "rhinoceros-hided." Briggs, *Age of Improvement*, 522; Armytage, "W. E. Forster," 209.

[16]See the London *Times*, 19 December 1857, 8. The group responsible for this effort had apparently been working toward such an end since 1853 at the least, and

James Lorimer published his *Political Progress Not Necessarily Democratic,* a widely-read book conceding the necessity of universal suffrage, but calling for the assignment of multiple votes to those with higher social standing.[17] In 1859 Thomas Hare's *A Treatise on the Election of Representatives, Parliamentary and Municipal* appeared. Hare detailed an enormously involved plan for circumventing the tyranny of the majority by permitting unsuccessful or superfluous supporters of particular candidates to have their votes reapplied to their contingent choices in other contests and locales.[18] In 1860 Henry Fawcett published a simplification and revision of Hare's plan.[19] The most prominent of all such liberal would-be bulwark builders was, however, J. S. Mill, who between 1859 and 1861 advanced, in various places, in addition to appreciations of Hare's plan, his own suggestions for a literacy and numeracy requirement, a poll tax, an exclusion of all on relief or not paying taxes, the

---

included by the later date such luminaries as the Chief Justices of England and Ireland; the president of the College of Physicians; various deans, both academic and religious; the vice-chancellors of Oxford and Cambridge; the president of the Royal Society; the Queen's surgeon; professors of almost every description; the headmasters of Harrow, Eton, Winchester and Rugby; and such assorted eminences as W. J. Hooker, John Scott Russell, Charles Babbage, John Ruskin, J. A. Froude, J. Symonds, William Hazlitt, and also William Bateson's father.

[17]Lorimer, *Political Progress Not Necessarily Democratic; Or, Relative Equality the True Foundation of Liberty* (London and Edinburgh: Williams and Hargate, 1857). J. S. Mill critiques this work in his "Recent Writers on Reform," *Essays on Politics and Society,* vol. 2, 352-58.

[18]Hare, *A Treatise on the Election of Representatives, Parliamentary and Municipal* (London: Longman, Brown, Green, Longmans, & Roberts, 1859), is also discussed in Mill's "Recent Writers on Reform," 358-70. An informative discussion of Mill's appreciation and advocacy of Hare's general idea appears in Paul B. Kern, "Universal Suffrage without Democracy: Thomas Hare and John Stuart Mill," *Review of Politics* 34 (July 1972): 306-22. See also Jon Roper, *Democracy and Its Critics: Anglo-American Democratic Thought in the Nineteenth Century* (London: Unwin Hyman, 1989), 150-51, 154; R. J. Halliday, *John Stuart Mill* (London: Allen and Unwin, 1976), 135; Alexander Brady, "Introduction," to Mill, *Essays on Politics and Society,* vol. 1, xxxiv-xxxv.

[19]Fawcett, *Mr. Hare's Reform Bill Simplified and Explained* (Westminster: Printed by T. Brettnell, 1860).

application of a test similar to the entrance examinations at Oxford and Cambridge, and an occupational scale whereby one vote would be accorded to unskilled laborers, two to skilled, three to foremen, three to four to farmers, manufacturers, and traders, and five to six to university graduates.[20] "It is an uphill race," he wrote to Fawcett in 1860, "and a race against time, for if the American form of democracy overtakes us first, the majority would no more relax their despotism than a single despot would."[21] Carlyle described the reaction of some conservatives in 1867. He asked, in seemingly rhetorical fashion, if there might just be a "small nucleus of invincible [*aristoi*] fighting for the Good Cause in their various wisest ways, and never ceasing or slackening till they die."[22]

---

[20] These suggestions appeared principally in Mill's *Thoughts on Parliamentary Reform* and in his *Considerations on Representative Government* (London: Parker, Son and Bourn, 1861). The following citations will refer to the reprinting of these works in Mill, *Essays on Politics and Society*, vol. 2. Discussion of literacy and numeracy requirements, a qualifying examination for multiple votes, and a sliding scale of votes by occupation appears in *Thoughts on Parliamentary Reform* on pp. 327-28, 325, and 324-25, respectively; discussion of literacy and numeracy requirements, a poll tax, exclusion of all on relief or not paying taxes, and the sliding scale appears in *Considerations on Representative Government* on pp. 470, 472, 472-73, and 475, respectively.

[21] Mill to Fawcett, 5 February 1860, in Francis E. Mineka and Dwight N. Lindley, eds., *The Later Letters of John Stuart Mill 1849-1873* (Toronto: University of Toronto Press,1972), 440.

It would be a longer and slower race than Mill and others of the time anticipated. As D. G. Wright observes,

> About one person in every twenty-four possessed a vote in 1832, about one in twelve in 1867 and one in seven in 1884. The 1884 Reform Act established democracy in principle rather than in practice. As late as 1911 only 29.7 percent of the total adult population of the United Kingdom were able to vote. Genuine manhood suffrage was delayed until 1918.

See Wright, *Democracy and Reform, 1815-1885* (London: Longman Group, 1970), 10.

[22] "Shooting Niagara," 330. Carlyle assured that "Anti-Anarchy is silently on the increase, at all moments. . . .To Anarchy, however million-headed, there is no victory possible." Indeed, he counseled, "Patience, silence, diligence, ye chosen of

Thus one great historical parallel obtaining between Galton's England and Plato's Athens was that for the privileged classes of both orders the gravest domestic challenge figured in their loss of political supremacy to their greatly more numerous social inferiors, a loss which had already been suffered by the Athenians and was increasingly imminent for Britons. Yet why should Francis Galton be associated with any of this? In the first place, it should be observed that the linking of English opponents of democracy with Plato and Platonism was a fairly familiar activity in the early twentieth century. In 1915 the English historian and political scientist Sir Ernest Barker characterized a broad swath of social commentators in this fashion, including Thomas Carlyle, John Ruskin, William Morris, Matthew Arnold, James Anthony Froude, W. R. Greg, W. E. H. Lecky, William Bateson, both the French and the English Positivists, and even the later Charles Dickens.[23] Simultaneously, American advocates of a more rigid social order who were demonstrating an informing appreciation of Plato ranged from Charles Warren, the lawyer and cofounder of the Immigration Restriction League, as early as his college days in 1889, to the biologist Charles W. Hargitt, who in 1913 invoked Plato's inspiration in advocating the examination of family pedigrees by college admissions boards, so as to restrict higher education to children of

---

the world! Slowly or fast in the course of time you will grow to a minority that can actually step forth (sword not yet drawn, but sword ready to be drawn), and say: 'Here we are, Sirs; we also are minded to *vote*, to all lengths as you may perceive. . . .'" (319, 336).

[23] Barker, *Political Thought in England: From Herbert Spencer to the Present Day* (New York: Henry Holt and Co., 1915). Discussion of the associated anti-democratic sentiments and Platonism (explicit or implicit) of Bateson, Carlyle, Ruskin, Arnold, Morris, the Positivists, Dickens, Froude, Greg, and Lecky appears on pp. 160-61, 188-97, 197, 197, 198-99, 199-201, 201, 201-02, 202, 202, respectively. Barker's own opinion seems fairly indicated by his assertions that "no believer in democracy would deny the great contention of Plato and Carlyle, that the aristocracy of the ablest and best should rule; but most believers in democracy would doubt whether such an aristocracy can be found by any other means than the free choice of all" (190).

the proper strains.[24] Thus in both England and America Plato was functioning as a patron saint and mentor for resistance to democracy.

Still, why should Robert Yerkes—and possibly other American hereditarians—have seen Francis Galton as one of Plato's anti-democratic disciples? Almost certainly this view did not derive from their acquaintance with the poem that Galton had submitted in competing for the Camden medal at Cambridge in 1843, which included the lines "Well may we burn to be citizens/ "of some state modeled after Plato's scheme" —not, at least, before 1914, the year in which the first volume of Pearson's biography was published. Nor could they have seen, until 1924, Pearson's statement that Galton believed that "racial improvement must depend upon the creation of a caste in each social class. . . ."[25] Clearly this perception of Galton as an enemy of democracy resulted in part from various statements

---

[24]See Charles Warren, "Plato's Republic with Reference Especially to the Functions of the State Therein and Its Relation with Some Modern Questions," Essay in Philosophy I, 1888-89, 9 February 1889, HUC 8888. 370. 1. 92. 2, Harvard University Archives. Cambridge, Massachusetts. As Barbara Miller Solomon has noted, in this paper Warren maintained "that the contemporary state needed social regulation to prevent the domination of the political majority," a theme which he developed further in his speech at commencement in June 1889, "The Failure of the Democratic Idea in City Government," Commencement Dissertation. See HUC 6889. 92, Harvard University Archives. Solomon, *Ancestors and Immigrants: A Changing New England Tradition* (Cambridge, MA: Harvard University Press, 1956), 100.

Charles Wesley Hargitt, the Curator of the Biological Museum and Director of the Zoölogical Laboratory at Syracuse University, praised Galton near the opening of his article as having "shown beyond reasonable doubt that genius [i.e., intelligence] follows the same laws which control other phases of development." After noting that he himself had been advocating eugenics since 1903, Hargitt turned directly to Plato, terming it "strangely significant that Plato conceived a similar ideal as the basis of his Republic. . . ." "Had he known as do we to-day the directing and controlling power of *heredity*," Hargitt deduced, "that Republic, instead of an *Utopia*, might have been an abiding *reality*, as glorious as the imperishable art and literature of its golden age!" From this point, as observed, Hargitt went on to advocate the use of Galton's scientific truths to sanctify the instituting of Plato's political ones. See Hargitt, "A Problem in Educational Eugenics," *Popular Science Monthly* 83 (October 1913): 358, 358-59, 359, 361-67.

[25]Pearson, *Life of Galton*, vol. 1, 177; Pearson, *Life of Galton*, vol. 2, 121.

in his own writings affording such an impression. For example, in 1911 the cereal baron and impresario of eugenics congresses Dr. J. H. Kellogg prominently quoted Galton's statement that *"the general standard of man is but little above the grade of trained idiocy"* (Kellogg's emphasis). Certainly all interested parties could find a wealth of such pronouncements in Galton's (and Pearson's) early twentieth-century addresses. Moreover, they could have found in one of his more famous earlier writings—a popular description of how eugenics could be established, published in 1873—the phrase that "the democratic feeling. . .[which] asserts that men are of equal value as social units, equally capable of voting, and the rest. . .is undeniably wrong and cannot last."[26]

More importantly, for some this view of Galton seems to have derived from critical scrutiny of the internal mechanisms of eugenics. The Columbia University sociologist Franklin H. Giddings, in a lecture delivered before one of his classes in April 1909, and before the readership of the *Popular Science Monthly* in July 1909, quoted from the English social analyst William Mallock's *Aristocracy and Evolution* (1898):

> The human race progresses because and when the strongest human powers and the highest human faculties lead it; such powers and faculties are embodied in and monopolized by a minority of exceptional men; these men

---

[26]Kellogg, "Tendencies Toward Race Degeneracy," *New York Medical Journal* 94 (2 September 1911): 461. This article, which was continued in the next number of the journal, constituted the text of an address which Kellogg had delivered before the Connecticut State Conference of Charities and Correction on 25 April 1911.

See "The Possible Improvement of the Human Breed under the Existing Conditions of Law and Sentiment," *Nature* 64 (31 October 1901): 659-65 as one example of Galton's early twentieth-century publications demonstrating his attitude toward the lower classes, especially p. 661, where he flatly declared that "the brains of the nation lie in the higher of our classes." This paper, a clarion call for the formation of a eugenics movement, was widely commented upon in the United States, and was reprinted in the *Annual Report of the Board of Regents of the Smithsonian Institution* (1901): 523-38, and in *Popular Science Monthly* 60 (January 1902): 218-33.

Galton, "Hereditary Improvement," *Fraser's Magazine*, January 1873, 127.

enable the majority to progress, only on condition that the
majority submit themselves to their control.

Immediately after citing this statement Giddings declared that
"no student of social evolution would be less likely to dispute these
propositions than Mr. Francis Galton, who, in fact, in his studies of
natural inheritance and hereditary genius, has done more than any
other investigator to establish them on a broad instructive basis."[27]

---

[27] Giddings, "Darwinism in the Theory of Social Evolution," *Popular Science Monthly*
75 (July 1909): 81.

CHAPTER FIVE: POSSIBLE INDICATORS
OF POLITICAL PURPOSE

THIS COURSE TAKEN by Giddings, of identifying anti-democratic purposes in Galton's eugenics formulations, seems to have been a tack taken by Yerkes and other hereditarian psychologists as well. Certainly of all Americans they were among those best equipped to comprehend such theoretical constructions, and to detect therein points at which Galton may have diverged from seemingly more intrinsically obvious lines of development into more ulteriorly indicated sectors and structurings. Accordingly we should inquire as to what features they could have perceived in Galton's development of eugenics – particularly in the theories of heredity which figured as carefully fashioned supports for his structure – that could have induced them to appreciate him as a politicizer of science generally, and more specifically as an opponent of democratic processes.

There are numerous possible features of this sort, appearing either in the forty-five-year-long course of Galton's development of eugenics or in the resulting doctrines themselves, that could have served to encourage such a view. Of these, nine possible candidates will be considered here.

1. *The nature of Galton's heredity tenets.*

For present purposes these four constituents of his system can be indicated fairly simply. The first and basal tenet, indeed the *sine qua non* of his eugenics, was that heredity is hard (this is, again, our modern term), which is to say that the processes of hereditary transmission are virtually impermeable to all outside influences and agencies: to all manner of environmental factors; to the incorporation of changes wrought by individual parents on the "ancestral" character of their own minds and bodies; and finally to such putative influences as birthmarking in all its various forms such as maternal impressions and telegony. Thus Galton was explicitly repudiating the doctrine of the transmission of acquired characters and much else besides. The second Galtonian tenet, which he came to call his Ancestral Law, contended that, when reduced to the utter average, the hereditary input of each ancestral generation is essentially halved as we look back into the past. Thus (Galton's arithmetic here was to be refined on several occasions) the parents of an individual contribute $\frac{1}{4}$ of his or her genetic makeup, the grandparents $\frac{1}{8}$, the great grandparents $\frac{1}{16}$, and so on. The third tenet, that of regression to the mean, held that, on average, the offspring of exceptional individuals will, rather than duplicate or further extend the aberration, regress toward the racial (or ancestral) average for that quality: exceptional parents bear less exceptional children. The fourth tenet maintained that evolution is discontinuous, that it proceeds not through the almost infinitesimally small gradations that characterized Darwin's scheme, but through the sudden emergence of rather radically new qualities, essentially through what would later be termed mutations.

These were remarkable contentions: some for their largely aleatory accuracy; most for the fact that almost no one at the time in which Galton had propounded them had either subscribed to them or conceived of them; and most as well for their seemingly superfluous character. Why did Galton, almost entirely uniquely, insist on the necessity of hard heredity? Darwin had hardly done so, and was to go on to give an ever-enlarged role in his system to soft-heredity processes, especially use-inheritance (the idea that protracted employment of a given organ both strengthened the organ and made it easier for one's offspring to effect exceptional development of the same organ—that blacksmiths' and scholars' sons could more easily develop large biceps and powerful intellects, respectively).[1] Why did Galton make so much of his Ancestral Law,

---

[1] For some of the modern commentary on Darwin's evolving position on soft heredity, see Robert C. Olby, *Origins of Mendelism* (New York: Schocken Books, 1966), 57, 60-62; Gerald L. Geison, "Darwin and Heredity: The Evolution of His Hypothesis of Pangenesis," *Journal of the History of Medicine and Allied Sciences* 24 (October 1969): 379, 388-91, 410-11; Peter J. Vorzimmer, "Darwin's 'Lamarckism' and the 'Flat-Fish Controversy' (1863-1871)," *Lychnos* 1969-1970: 121-70; Leonidas Iakovelovich Blacher, *The Problems of the Inheritance of Acquired Characters: A History of the A Priori and Empirical Methods Used to Find a Solution*, trans. Katy McKinin and Noel Hess, English translation ed. Frederick B. Churchill (New Delhi: Amerind Publishing Co., for the Smithsonian Institution Libraries and the National Science Foundation, Washington, D. C., 1982; originally published Moscow: Nauka

a starkly abecedarian generalization first elaborated at least a century and a half before, and of no use whatsoever for the description, much less the prediction, of typical hereditary results?[2]

Publishers, 1971), 32, 35, 36, 41-42 (note that Blacher had preferred that his name be transliterated in this form but that other editions or references to him might take the form employed by the Library of Congress, i.e., Blâikher); Richard W. Burkhardt, Jr., "Closing the Door on Lord Morton's Mare: The Rise and Fall of Telegony," *Studies in the History of Biology* 3 (1979), 4, 6-8, 11; Ernst Mayr, *The Growth of Biological Thought: Diversity, Evolution, and Inheritance* (Cambridge, MA: Belknap Press of Harvard University Press, 1982), 689-94; Frederick B. Churchill, "From Hereditary Theory to *Vererbung*: The Transmission Problem, 1850-1915," *Isis* 78 (September 1987): 343-46.

For some of more important participants in the earlier discussion of the same issue, see [E. Ray Lankester,] "Darwin Versus Lamarck" ("Abstract of a Lecture Delivered at the London Institution, Finsbury Circus, On February 14, 1889, by Prof. Ray Lankester, LL.D., F.R.S."), *Nature* 39 (28 February 1889): 428-29; "Opening Address by Prof. Sir William Turner, M. B., LL.D., FR.SS.L. & E., President of the Section" (of Anthropology, of the British Association), *Nature* 40 (26 September 1889): 531; Herbert Spencer, letter to editor, *Nature* 41 (6 March 1890): 414-15; E. Ray Lankester, letter to editor, *Nature* 41 (6 March 1890): 415-16; E. Ray Lankester, letter to editor, *Nature* 41 (27 March 1890): 486-88; Alfred R. Wallace, "Are Individually Acquired Characters Inherited? II.," *Fortnightly Review*, May 1893, 663-64; J. Cossar Ewart, "Opening Address by Prof. J. Cossar Ewart, M.D., F.R.S., President of the Section" (of Zoology, of the British Association), *Nature* 64 (12 September 1901): 482-84, 487; J. Arthur Thomson, *Heredity* (London: John Murray, 1908), 144-50; T. H. Morgan, "For Darwin," *Popular Science Monthly* 74 (April 1909): 373; R. Medola, "A Contribution to the History of Evolution," *Nature* 85 (5 January 1911): 297-98; W. T. Thiselton-Dyer, letter to editor, *Nature* 85 (19 January 1911): 371; John W. Judd, letter to editor, *Nature* 85 (26 January 1911): 405-06; E. A. Parkyn, letter to editor, *Nature* 85 (9 February 1911): 474; John W. Judd, letter to editor, *Nature* 85 (9 February 1911): 474-75; P. Kropotkin, "Inheritance of Acquired Characters: Theoretical Difficulties," *Nineteenth Century and After*, March 1912, 512-16.

These citations from the earlier period indicate simply that American hereditarians of the early twentieth century, who recognized the political implications of hard heredity as well as anyone, had ample opportunity to appreciate Galton as a far firmer proponent of this doctrine than Darwin had been.

[2] For earlier articulations of the principle that the average ancestral contribution to an individual's heredity halves by each generation working backward, see William Wollaston, *The Religion of Nature Delineated* (London: Sam. Palmer, 1724), sect. VIII, "Truths concerning Families and Relations" (which extends over pp.154-167), 161. Wollaston cites a number of classical antecedents, and expresses the principle abstractly: "For let C be the son of A and B, D the son of C, E of D, F of E: and let

Moreover, what purpose could this abstraction have served, given that anyone of the time could see reversion — the reappearance, sometimes in dominant fashion, of ancestral qualities not seen in several generations — to be an everyday staple of heredity? Why his insistence on discontinuity (saltations) in evolution, when Darwin, his ostensible guide in such matters, had made gradualism a necessary article of faith in his system? Were Galton to have proceeded in accordance with the Darwinian apprehension of heredity which he professed to revere, he could easily have shaped a program for directed evolution that featured none of these tenets. What purposes could they have been seen as serving for him?

If the imminent rise and possible dominance of the lower classes could be seen as representing a massive threat to middle and upper class prospects in the 1860s and beyond, Galton's heredity tenets could be appreciated as so directly apposite a response that it would be hard to suggest accidentality. Hard heredity fitted this function most vitally. Without it Galtonian eugenics would be neither conceivable nor required. If the tenets of soft heredity were accurate and acquired characters could be transmitted to succeeding generations, or if offspring could be shaped to order through the exercise of "prenatal culture" (deliberately inculcated maternal

---

the *relation* of C to A and B be as [one]: then the *relation* D to A and B is but the half of that, which C bears to them. By proceeding after the same manner it will be found, that the *relation* of E to A and B is ¼ (or half of the half), of F $^1/_8$: and so on. . . ." Note that Wollaston's book was first printed in 1722, but only in a version which he had not authorized, "with many mistakes and *errata*."

See also Charles White, *An Account of the Regular Gradation in Men, and in Different Animals and Vegetables; and from the Former to the Latter* ("Read to the Literary and Philosophical Society of Manchester, At Different Meetings, in the year 1795") (London: Printed for C. Dilly, In the Poultry, 1799), 117. White's contribution comprises a table specifying names for the various combinations of whites and blacks up to four generations back, ergo to sixteenths, e. g., the offspring of a "White and Mestize" ("Mestize" being $^7/_8$ white and $^1/_8$ black) is "$^{15}/_{16}$ White and $^1/_{16}$ Black," and "is denominated" "A (reputed) White." On pp. 145-46, White with the assistance of "a person skilled in numbers" attempted a very early population study, calculating that if a colony composed of equal numbers of blacks and whites bred indiscriminately, in 65 years the numbers of blacks, whites and mulattoes would be equal; in 91 years the whites and blacks would each be one-tenth of the whole number, and in three centuries they would each be one one-hundredth.

impressions), there need be little concern for defective strains or, even, heredity itself. Instead great effort should be directed toward environmental amelioration: potentially degenerative social influences should be eradicated; opportunities for unbounded self-improvement, along with encouragement for remaking fetuses *in utero*, should be extended to every potential breeder; and no family, strain or race should be considered to stand outside the pale of perfectibility. Conversely, of course, if a general subscription to fixed heredity could be effected, corollaries of an entirely antithetical sort could be advanced: effectively that, because breeding is everything, all environmental palliatives, all social leveling, all efforts toward democratization, indeed most forms of charity, should be suspended forthwith. Moreover, because a natural biological sifting, one resulting in a propitious social stratification, had long been in progress, those already on top—and at the bottom—should be helped to stay there.

Galton's Ancestral Law and his theory of regression to the mean could be seen as theoretical inducements for moving even further in the direction of oligarchy. The former tenet, by predicating that, on the average, each parent could determine only one-eighth of his or her child's heredity,[3] and the latter, by contending that, on the average, the offspring of exceptional types must revert toward their ancestral mean, could both militate against social mobility. Because

---

[3] In a letter to the editor of *Nature* Galton once made this principle explicit:

> The exceptional quality of the father is only one of four elements that contribute in apparently equal shares to determine the position of the genetic focus. The other three are (1) The quality of the mother, (2) That of the paternal ancestry, (3) That of the maternal ancestry. . . . Consequently the exceptional quality of the father, considered apart from his ancestry, is not likely to raise the position of the joint genetic focus of himself and the mother by more than a quarter of its amount.

Note that one-fourth, *vice* one-eighth, was the proportion which Galton was according to each parent at that particular point in the evolution of his Ancestral Law, and also that Galton went on in this same letter to ascribe similar exceptionality-diminishing effects to regression, naturally enough. See Galton, "On the Probability that the Son of a Very Highly-Gifted Father Will Be No Less Gifted," *Nature* 65 (28 November 1901): 79.

both tenets indicated the overwhelming importance of ancestry, both pointed up the racial (phylogenetic) impropriety not only of conducting searches for superior types in the lower orders, but also of allowing such "sports" to rise once having manifested themselves. If such exceptional individuals are generally incapable of producing equally exceptional progeny, if the imprint of their inferior ancestry is forever on their issue, these should hardly be situated where they might interbreed with dependably superior stocks, and thus dilute their quality. Moreover, conversely focus on ancestry could serve to discourage the demotion of inferior individuals of superior stock as well, inasmuch as their ancestral superiority could be expected to raise the offspring of such negative exceptions. Thus Galton's Ancestral Law and his concept of regression to the mean could be seen as working in tandem, and consistently, in favor of the superior social classes and against the lower.

Yet, it could also have been perceived, such a concept as regression to the mean could well have discouraged superior persons on any social level from attempting to improve the race through carefully selected marriages. Thus the ulterior service performed by Galton's remaining tenet: his doctrine of discontinuous evolution could be regarded simply as providing a necessary counterforce to his law of regression. If seeming evolutionary advances must generally be succeeded by next-generation regressions to the mean, evolution, both natural and directed, in order to work at all, must depend upon occasional broad leaps to new and stable positions.[4]

---

[4]For useful discussions of any or all of these four heredity tenets, consult W.K. Brooks, "Francis Galton on the Persistency of Type," *American Journal of Psychology* 1 (November 1887): 173-79; "A New Law of Heredity," *Nature* 56 (8 July 1897): 235-37; J. McKeen Cattell, review of "The Average Contribution of Each Several Ancestor to the Total Heritage of the Offspring" by Francis Galton, *Psychological Review* 4 (November 1897): 676-77; Galton, *Memories*, 287-309; Thomson, *Heredity*, 309-35; H.S. Jennings, "Experimental Evidence of the Effectiveness of Selection," *American Naturalist* 44 (March 1910): 140; Pearson, *Life of Galton*, vols. 2, 3A (Cambridge: Cambridge University Press, 1930), 3B, passim; J.S. Wilkie, "Galton's Contribution to the Theory of Evolution with Special Reference to His Use of Models and Metaphors," *Annals of Science* 11 (September 1956): 194-205; R.G. Swinburne, "Galton's Law — Formulation and Development," *Annals of Science* 21

2. *The sequential position of Galton's introduction of his heredity tenets.*

Here the concern is Galton's warrant for making such claims. Did they originate as the product of scientific inquiry (disinterested or not), or as *a priori* pronouncements, still needing to be proven (and thus possibly as dictated responses to the political concerns of his time)? Clearly, such an issue mattered greatly to Karl Pearson, who took considerable pains to represent all of Galton's heredity tenets and eugenics itself as the laboriously uncovered products of a patient Baconian induction.[5] Instead, as Pearson well knew,

---

(March 1965): 15-31; Ruth Schwartz Cowan, "Sir Francis Galton," dissertation, 1-44, 87-269; Olby, *Origins of Mendelism*, 68-83; Geison, "Pangenesis," 378-79; P. Froggatt and N.C. Nevin, "The 'Law of Ancestral Heredity' and the Mendelian-Ancestrian Controversy in England, 1889-1906," *Journal of Medical Genetics* 8 (March 1971): 1-36; Froggatt and Nevin, "Galton's 'Law of Ancestral Heredity': Its Influence on the Early Development of Human Genetics," *History of Science* 10 (1971): 1-27; Ruth Schwartz Cowan, "Galton and the Continuity of Germ-Plasm," 181-86; William B. Provine, *The Origins of Theoretical Population Genetics* (Chicago: University of Chicago Press, 1971), 14-24, 33-34, 51-55, 179-87; Ruth Schwartz Cowan, "Francis Galton's Contributions to Genetics," *Journal of the History of Biology* 5 (Fall 1972): 389-412; B.J. Norton, "The Biometric Defense of Darwinism," *Journal of the History of Biology* 6 (Fall 1973): 290-93; Forrest, *Francis Galton*, 187-206; Ruth Schwartz Cowan, "Nature and Nurture," 133-208; Bernard Norton, "Psychologists and Class," in ed. Charles Webster, *Biology, Medicine and Society 1840-1940* (Cambridge: Cambridge University Press, 1981), 289-314; Mayr, *Growth of Biological Thought*, 695, 784-85, 891; Raquel Alvarez Peláez, "Las Fuentes Francesas de La Eugenesia de Galton," *Asclepio* 37 (1985): 165-81.

[5] This exercise in chronological reorganization began for Pearson in 1911 when he composed Galton's obituary. Here he described Galton's "labours to make anthropometry. . . an exact science," and "his discovery that new types of analysis are wanted to replace mathematical function in biological studies," and, "lastly, his advocacy of Eugenics" as "successive steps in a continuous ascent" — despite the fact that Galton first advocated eugenics in 1865, decades before these other activities, which were meant to give scientific ballast to eugenics. Here Pearson also claimed that Galton had "originally stated" his Ancestral Law in *Natural Inheritance*, thus in 1889, twenty-four years after its actual first appearance. See Pearson, "Francis Galton," 441, 443. In 1922, in his *Centenary Appreciation* of Galton, Pearson celebrated those brilliant English Victorian scientists whose "method was essentially Baconian": Darwin, Lyell, Hooker, Faraday, Maxwell, Rayleigh, Kelvin, and Clifford "accumulated facts and allowed the trained imagination to play upon them." Likewise Galton, who, according to this account, conceived of hard heredity

Galton was bound to all of his tenets, and presented them as facts of heredity, well before conducting any investigation to confirm their veracity. Indeed, all were present—albeit some in embryonic form—at the creation: Galton enunciated all of his heredity tenets in "Hereditary Talent and Character," all except for the doctrine of discontinuous evolution, which made its first appearance in *Hereditary Genius*, four years later.[6] Naturally, it would hardly do for Pearson to admit that Galton knew from the outset the laws which he and his eugenics needed, declared their existence by *fiat*, and then set out to discover the evidence required to prove them. Yet any relatively cognizant reader could deduce as much through a reading of "Hereditary Talent and Character."

3. *The nature of Galton's defense of his heredity tenets.*

　　Only two of Galton's tenets met with any general opposition

---

only after his inability to confirm, through a course of patient experimentation, Darwin's hypothesis of pangenesis, hence circa 1872. Moreover Galton is represented as coming to his basic premise that heredity prevails over environment in the making of ontogenetic intelligence only circa 1875, after a careful investigation of twins. See Pearson, *Centenary*, 7, 15, 17. Somehow this consideration of Galton could be written even after the first volume of Pearson's biography of Galton had been published—in 1914—with its explicit observation that Darwin's "Baconian method was not Francis Galton's." Perhaps by 1922 Pearson had come to feel that this view of Galton's method was not serviceable. See Pearson, *Life of Galton*, vol. 1, 58. Finally, in 1924, when required to chronicle Galton's opening elaboration of eugenics (i.e., "Hereditary Talent and Character"), and apparently wanting to secure the earliest priority for Galton in these areas, Pearson silently reversed himself, explaining that Galton had first advocated eugenics, and conceived of hard heredity and the Ancestral Law, in 1865. More than this, he remarked it to be "singular how this foundation stone. . .has been disregarded even by some of his professed followers." See Pearson, *Life of Galton*, vol. 2, 75-87, and for the observations and remarks specified above, 77-78, 81-82, 84, and 75, respectively.

[6] "Hereditary Talent and Character," "Second Paper," 321-22 for Galton's presentation of his hard heredity argument; "Part I," 158, "Second Paper," 319, for statements resembling and anticipating his formal articulation of the process of regression; "Second Paper," 326-27, for his first statement of the basic principle involved in the Ancestral Law; *Hereditary Genius*, 368-70 for two analogies expressing his theory of discontinuous evolution.

(up until his transfer of doctrinal authority to Pearson) and these he defended with a fierce and sometimes calculating determination. When his doctrine of discontinuous evolution came under fire in the early 1890s, he stood his ground openly, rallying all the support he could muster.[7] Conversely, his defense of hard heredity, which was mired in difficulty almost from the first, required a much more sustained and sometimes clandestine effort. This remarkable campaign needs to be described at some length, as it could well have illustrated the vital importance of hard heredity for Galton, and the intensity of his commitment thereto.

Galton's problems with hard heredity dated from his first awareness of Darwin's "provisional hypothesis" of pangenesis. If Galton received this information as early as two years before his cousin's official introduction of the theory in *The Variation of Animals and Plants* (1868) we should have reason to conjecture that this could have been the cause of the second, more devastating, breakdown that afflicted him in 1866.[8] (Such speculation may not be altogether

---

[7] For Galton's further elaborations and defenses of discontinuous evolution see Galton, [President's address,] *Journal of the Anthropological Institute* 15 (1886): 495-96; Galton, *Natural Inheritance*, (London: Macmillan and Co., 1889): 18-34; Galton, "The Patterns in Thumb and Finger Marks. — On Their Arrangement into Naturally Distinct Classes, the Permanence of the Papillary Ridges that Make Them, and the Resemblance of Their Classes to Ordinary Genera," *Philosophical Transactions of the Royal Society of London* (hereafter *Philosophical Transactions*) 182B (1891): 20-23; Galton, "Prefatory Chapter to the Edition of 1892," *Hereditary Genius*, 2d ed., xviii-xix; Galton, "Discontinuity in Evolution," *Mind* 19 n.s. (July 1894): 362-72. His most significant support came from William Bateson, *Materials for the Study of Variation, Treated with Especial Regard to Discontinuity in the Origin of Species* (London: Macmillan and Co., 1894). See Provine, *Origins*, generally, for discussions of Huxley's possible influence on, and Bateson's support of, Galton's theory of discontinuous evolution. See also Cowan, "Nature and Nurture," 192-93, for an excellent discussion of the attitudes which Galton demonstrated in his study of fingerprints.

[8] For discussions of this breakdown, see Galton, *Memories*, 154-55, 161, 215; Forrest, *Francis Galton*, 85-88; Blacker, *Eugenics*, 36-37; Slater, "Galton's Heritage," 99. Galton considered this four-year-long affliction to have been one of the most important events of his life, and notes that he was so severely impaired that he would have been "playing with death" had he continued to hold his position of General Secretary of the British Association in this period (see *Memories*, 215).

farfetched, considering that Darwin had submitted a lengthy elaboration of the theory to Thomas Huxley by that time, and that Galton and Huxley were in frequent contact in the same period, as fellow owners and editors of the *Reader*, through regular communications with regard to the program for the British Association, which both were responsible for preparing, and a result of their mutual attendance at various clubs and Royal Society functions.[9]) Whenever it was that Galton learned of pangenesis, he could be expected to have suffered greatly. For here he was, embarked on what would be his life's great work, having just proclaimed a system for perfecting humanity, and—if the news reached him only after the publication of Darwin's book—well in the throes of assembling *Hereditary Genius*, only to discover that hard heredity, the entirely indispensable basis of his system, had been unceremoniously (and surely unconsciously) plowed under by Darwin's theory. Pangenesis had been designed to account for all the generative processes from sexual and asexual reproduction, and the regenera-

---

Hence it is intriguingly curious that Pearson could write a 1,300-page biography of Galton discussing his lesser illnesses in at least eleven separate places and entirely avoid mention of this crucial and long-extended one. (Pearson does, seemingly inadvertently, include one reference to the disorder in reproducing the entry which Louisa Galton made in her *Record* for 1869, which included the notation "Frank in good health and able to dine out again." See Pearson, *Life of Galton*, vol. 2, 88n.) Perhaps Pearson's reluctance to refer to this episode may have stemmed from an unwillingness to explain that Galton was suffering from severe mental strain and obsessional tendencies at the same time that he was composing *Hereditary Genius* (at great risk to his mental and physical health).

[9] For Darwin's discussion of pangenesis with Huxley, see Francis Darwin, ed., *The Life and Letters of Charles Darwin*, 3 vols. (London: John Murray, 1887), vol. 2, 227-28; Leonard Huxley, *Life and Letters of Thomas Henry Huxley*, 2 vols. (London: Macmillan and Co., 1900), vol. 1, 268. For particularly useful letters on pangenesis see Darwin to Huxley, 12 January 1867, Scientific and General Correspondence, vol. 5, pp. 235b-236a, Thomas Henry Huxley Papers, College Archives, Imperial College of Science, Technology and Medicine, London, England, hereafter referred to as Huxley Papers. For an indication that Galton was in contact with Huxley in this period, see Galton to Huxley, 15 August 1866, regarding officers for Section D of the British Association, Scientific and General Correspondence, vol. 17, G-HAN, p. 5, Huxley Papers.

tion of lost parts in plants and some animals, to gemmation, the process enabling severed pieces of some genera to grow into complete plants. Thus it provided a mechanical model for soft heredity. By theorizing generation to occur in the cells themselves, and these cells to produce "gemmules" which reflected their parent cells' changing conditions and character, and these gemmules to be stored and passed on to future generations, Darwin had rather plainly, if entirely hypothetically, explained the transmission of acquired characters.[10]

Thus, it would seem, if Galton meant to keep his nascent eugenics enterprise afloat he would have had three choices. He could embrace pangenesis, so as to emphasize those of its elements that might be turned to his own ends, or he could work to destroy the theory, either by disproving it or by advancing a contradictory one of his own.[11] Yet all of these courses should have seemed perilous, not only because Darwin was his cousin in a day when family loyalty mattered greatly, but also because Darwin figured as a world authority in natural history, a field in which he himself was largely unread.

Nonetheless, Galton may have elected to attempt all three

---

[10] Darwin's presentation of pangenesis appeared in Chapter 27, "Provisional Hypothesis of Pangenesis," pp. 357-404, in vol. 2 of Darwin, *Variation*.

[11] Note that the idea speculated here, that Galton might secretly have intended to oppose pangenesis from the first, constitutes a wholly unique interpretation of the famous pangenesis affair. Yet such a reading does have the virtues of according more faithfully with Galton's obvious attachment to (and possible dependence upon) hard heredity, and of helping to explain one feature in the affair which has seemed incomprehensible to all who have written on the episode, namely Darwin's severely petulant behavior. Clearly Darwin felt that Galton had betrayed him, as might be perceived in Galton's sensed need to assure Darwin on 12 May 1871 that a particularly dismissive letter written by Lionel S. Beale to *Nature* on the subject of pangenesis had not resulted from Beale's having spoken with, or been influenced by, himself. "I do not know him;" Galton insisted, "at least, I have, perhaps twice only, had occasion to converse with him,—and what he says, certainly does not express my own opinion as expressed elsewhere and to others. I should not feel easy, if I did not disavow all share in it to you." See Pearson, *Life of Galton*, vol. 2, 162. See also Lionel S. Beale, "Pangenesis," letter to editor, *Nature* 4 (11 May 1871): 25-26.

strategies almost simultaneously, albeit sometimes in intricately surreptitious fashion. First, he left all mention of hard heredity out of *Hereditary Genius*, although the doctrine had been presented as the bedrock basis of eugenics in "Hereditary Talent and Character." Instead he closed the book with its chapter discussing pangenesis. Subsequent developments indicate that Darwin accorded little attention to this discussion, for reasons suggested by its highly curious nature. Galton opened with the observations that the "'Provisional'" theory was based "on pure hypothesis and very largely on analogy," yet that, "whether it be true or not," it would be of "enormous service to those who inquire into heredity."[12] Such remarks, however variably approbative, were both justified and largely obvious, yet Galton proceeded to inject a series of ingenious but distractingly prosaic analogies of his own. This he did to explain how gemmules congregate to form cells of a given type for particular organs and next how the hereditary materials from two conjoining families might either work together or compete for dominance.[13] These analogies suffered from being practically impenetrable, however, and soon Galton was continuing on into even larger problems. He demonstrated a serious misapprehension of one detail of Darwin's evolutionary theory, claiming that the process could actually be reversed and a species could revert to an earlier point in its development and then follow down the line taken by a distant relative: that dogs, for example, could return to the common ancestor they share with bears and become bears themselves. Men could do the same with apes.[14] Next, in this process, ostensibly to discuss pangenesis, he interposed an entirely digressive pair of analogies introducing his own theory of discontinuous evolution, seemingly without recognizing the affront posed thereby

---

[12] P. 364.

[13] Pp. 364-68.

[14] Pp. 368-69. Note that Galton did not provide such examples, having merely stated that "A and B having both descended from C, the lines of descent might be remounted from A to C, and redescended from C to B" (369).

to Darwin's doctrine of gradual evolution.[15] From here he launched into a very confusing discussion of hereditary processes that woefully misappropriated the term "individual variation," giving it two separate and highly problematic meanings.[16] (Galton appears to have meant by the term the variation caused by inherited acquired characters but also, in another place, the acquisition of these characters by the parents ontogenetically.) Next he ventured into an algebraic illustration of how the theory of pangenesis could be used mathematically to calculate (in a fashion anticipating the operations of the biometrical school that he and Pearson later founded) exactly how heredity might unfold for any given family. (What would be required was complete data on all the phenotypic characters of every ancestor for many preceding generations !)[17] Darwin must have laid the chapter down in bemused exasperation.

Yet if Darwin had persisted patiently, especially through the area where Galton was treating his "individual variations," he might have detected his cousin's tendency to emphasize the importance of ancestry, even if in a most indirect fashion. There, for "the sake merely of a very simple numerical example," Galton had taken the course of assigning one-tenth of a given child's nature to the inheritance of acquired characters and the remainder to what he had generally been terming "unchanged inheritance."[18] The purpose of this particular exercise had been ostensibly to show that "extremely ancient" ancestors contribute extremely little to the child's makeup, but another result of his calculations was to suggest that (except for the 10 percent deriving from the putative transmission of acquired characters) only 9 percent of hereditary material could come from both parents combined, and that fully 81 percent derived from earlier ancestry. Galton would make nothing of this consideration

---

[15] Pp. 369-70.

[16] Pp. 370-75.

[17] For these definitions, see 370 and 373. The algebra appears on 371-72n. Galton's explanation that "the latent gemmules. . .admit of being determined from the patent characteristics of many previous generations. . . ." appears on 373.

[18] P. 371.

in this setting but precisely this idea would serve as the core of his counter-argumentation in subsequent publications. Moreover, in the present discussion Galton also gave some emphasis to a closely associated observation.[19] Darwin, in accounting for the phenomenon of reversion (e.g., the sudden reappearance of a great-grandfather's distinctive nose), had ruled that some proportion of the gemmules in each individual's system must be latent—passed on and not utilized—and Galton had recognized that if these unutilized determiners existed for each trait they must exist in enormous numbers, given the broadly fanning increase of ancestors into the past. Hence, his subsequent studies would show him to have reasoned, the patent determiners, those manifested (phenotypically) by the parents themselves, must be of statistically minuscule importance, and the latent ancestral pool vastly more important, to the determining of each prospective child's makeup. In his (however-risible) examination of Darwin's soft heredity theory Galton had seized upon some measure of proof, even if only that derived from the logic of numbers, for the overwhelming importance of ancestry and thus, to some extent, hard heredity itself.

Next Galton (perhaps shrewdly) enlisted Darwin's aid in conducting transfusion and breeding experiments with rabbits to test pangenesis. (These experiments would consist of transfusing the blood of rabbits of one variety into rabbits of another to see whether offspring would be affected when the recipients were bred with their own kind.)[20] Galton wrote to Darwin on December 11, 1869, asking if he could recommend someone to assist and advise him in procuring and breeding rabbits of "marked and assured breeds" for a "really curious experiment," yet without specifying its purpose, and Darwin, who was certain to guess for himself, snapped

---

[19] Galton's latency argument is developed in the first complete paragraph of 371 through the last complete paragraph of 373.

[20] See Forrest, *Francis Galton*, 102-05, for a clear description of the experiments. See also Pearson, *Life of Galton*, vol. 2, 157-77; Ruth Schwartz Cowan, "Sir Francis Galton," dissertation, 110-12, and C.B. Davenport, "Light Thrown by the Experimental Study of Heredity upon the Factors and Methods of Evolution," *American Naturalist* 46 (March 1912): 135.

at the bait. Possibly Galton drew Darwin into the project both to dispel general perceptions of disloyalty and to deflect Darwin's objections should he succeed in disproving pangenesis. Certainly Galton assured his cousin that he was altogether intent on confirming the theory, and all the ensuing correspondence from London to Down attested to his persisting enthusiasm for that goal, even in the face of his recurrent failure to establish pangenetic influence.[21]

Only if Darwin had been able to peruse Galton's excitedly negative comments in the margins of the sections of his copy of *Variation* expounding the transmission, *via* gemmules, of acquired characters, or had examined the final chapter of *Hereditary Genius* sufficiently closely, would he have been prepared for what came next. On March 30, 1871, Galton appeared before the Royal Society, and subsequently on the pages of its *Proceedings*, "negativing. . . beyond all doubt, the truth of the doctrine of Pangenesis," in view of the failure of his experiment to corroborate the theory.[22]

---

[21] For the text of Galton's 11 December 1869 letter to Darwin, see Pearson, *Life of Galton*, vol. 2, 157: and Cowan, "Sir Francis Galton," dissertation, 109, for an abridged version. For the ensuing correspondence, see Pearson, *Life of Galton*, vol. 2, 157-62.

[22] For the nature and some of the particulars of Galton's marginalia to his copy of *Variation*, see Cowan, "Sir Francis Galton," dissertation, 96-99. Cowan notes that Galton marked up the pangenesis chapter much more heavily than any other part of the work and that he took "extensive long-hand notes" as well. Cowan describes Galton's reaction to pangenesis as "immediate and favorable," yet she also describes him as contesting rather insistently various of Darwin's evidences for the transmission of acquired characters; indeed, she claims, "Galton had a skeptical reaction whenever Darwin wrote of [acquired characters]" (98). Thus, "in light of Galton's skepticism about the inheritance of acquired characters," she finds it "something of a surprise to discover that he reacted favorably to the theory of pangenesis which Darwin presented in the concluding chapters of the book" (99). Yet she finds most of this favorable reaction in the final chapter of *Hereditary Genius* rather than in Galton's marginalia or notes to *Variation* itself.

For Galton's appearance before the Royal Society, see Francis Galton, "Experiments in Pangenesis, by Breeding from Rabbits of a Pure Variety, into Whose Circulation Blood Taken from Other Varieties Had Previously Been Largely Transfused," *Proceedings of the Royal Society of London* (hereafter *Proceedings of the Royal Society*) 19 (16 June 1870 to 15 June 1871): 393-410. The statement quoted

Unmistakably Darwin felt betrayed by this move on Galton's part, as can be seen in the extremely rare public response that he published in *Nature* on April 27, 1871. Here he repaid Galton by pretending neither to have participated in Galton's experiments nor ever to have made the claim which they had been conducted to test — that his hypothetical gemmules are transported in the blood.[23]

Galton, for his part, having been so forcibly unhorsed, reverted to what appears to have been more subterfuge, professing to acquiesce to Darwin's superior wisdom, and humbly resuming the experiments (which extended for another eighteen months) along with the assurances, made to Darwin, of his eagerness for success in the effort.[24] Yet even at the same time, in June 1872, he published another article propounding, although in densely analogical fashion, his directly opposed theories on latency and hard heredity.[25] A year

appears on p. 395. For Pearson's discussion of the immediate prelude to, and aftermath of, Galton's appearance before the Society, see Pearson, *Life of Galton*, vol. 2, 162.

[23] For Darwin's letter see 'Pangenesis,' letter to editor, *Nature* 4 (27 April 1871): 502. For Pearson's discussion of Darwin's behavior, see *Life of Galton*, vol. 2, 163-64, 165; for Cowan's see "Sir Francis Galton," dissertation, 112-13; for Forrest's, see *Francis Galton*, 103.

[24] For Galton's published response to Darwin's letter, see "Pangenesis," letter to editor, *Nature* 4 (4 May 1871): 5-6, or see it reproduced in Pearson, *Life of Galton*, vol. 2, 164-65. For Pearson's feelings on Galton's response, "one of the finest things he ever wrote in his life," see ibid., 165-66. For a similar reaction, see Cowan, "Sir Francis Galton," dissertation, 113-15. See also Forrest, *Francis Galton*, 104-05.

For Galton's resumption of the experiments, see Pearson, *Life of Galton*, vol. 2, 165-69, 174-77, 181, and Cowan, "Sir Francis Galton," dissertation, 115-16. Pearson says in one place (165) that the resumed experiments "went on for another three years at least!," and in another (177) that they had persisted "for two or more years," but Cowan seems correct in determining that they extended over the course of eighteen months (115).

[25] Galton, "On Blood Relationship," *Proceedings of the Royal Society* 20 (13 June 1872): 394-402. This paper was received by the Royal Society on May 7, 1872. For useful discussions of this article, see Cowan, "Sir Francis Galton," dissertation, 123-31; Pearson, *Life of Galton*, vol. 2, 169-74. For Galton-Darwin correspondence regarding this paper, see ibid., 169. See also for a helpful if less extended treatment of the paper, Forrest, *Francis Galton*, 105-06.

later he again presented these theories in the opening paragraphs of his first article responding to de Candolle.[26] Then in 1875, 1876, and 1877 he published three more papers advancing further in the same direction, over the increasingly animated objections expressed by Darwin upon previewing them.[27] Thus, including his *Hereditary Genius* chapter and his pangenesis confutation, Galton published during a period of eight years seven statements which attempted either to invalidate pangenesis or to promote his own theory, even while, through much of this time, representing himself before Darwin as a loyal and enduring supporter of the "provisional hypothesis." The widely-noted pangenesis affair, along with its aftermath, could well have been perceived as demonstrating how much hard heredity—and the contingent survival of eugenics—had mattered to Galton.

---

[26] Galton, "On the Causes," 345-46.

[27] "A Theory of Heredity," *Contemporary Review* 27 (December 1875): 80-95; "A Theory of Heredity" (which Galton describes as a version of the previous paper which he has revised "considerably" and amended "in many particulars"), *Journal of the Anthropological Institute of Great Britain and Ireland* 5 (January 1876): 329-48; "Typical Laws of Heredity," *Notices of the Proceedings of the Meetings of the Members of the Royal Institution of Great Britain with Abstracts of the Discourses Delivered at the Evening Meetings* 8 (1875-1878): 282-301. The last paper, read on 9 February 1877, is largely concerned with Galton's attempts to attribute causes to Quetelet's observations regarding deviation from an average, through his use of analogies, graphs, physical models, his famous experiments with sweetpeas, and, in his appendix thereto, the use of algebraic equations. Here Galton is refining his ideas on regression, which at this point he was terming "reversion." For insightful discussion of the breakthrough paper and the work leading up to it, see Cowan, "Sir Francis Galton," dissertation, 169-79; Cowan, "Francis Galton's Statistical Ideas: The Influence of Genetics," *Isis* 63 (December 1972): 512-21, and Donald MacKenzie, *Statistics in Britain, 1865-1930: The Social Construction of Scientific Knowledge* (Edinburgh: Edinburgh University Press, 1981), 60-63. See also Forrest, *Francis Galton*, 187-88; Pearson, *Life of Galton*, vol. 3A, 6-11; and Wilkie, "Galton's Use of Models and Metaphors," even though it makes no reference to "Typical Laws" itself.

For Darwin's objections, see Pearson, *Life of Galton*, vol. 2, 182-84, 187-91.

4. *Occasional breaks in Galton's demonstrated attachment to his heredity tenets.*

If Galton manifested a determined commitment to his heredity tenets, especially hard heredity, through most of his career, nevertheless there appear to have been some occasions on which he deemed it advisable to behave otherwise. One of these has already been discussed: when, after having made hard heredity so crucial a part of his inaugural exposition of eugenics, he failed to mention the doctrine in the major statement that came four years later. One possible reason for this has been suggested as well: because Darwin had published his theory of pangenesis in the preceding year, Galton may have been unwilling to oppose his highly-regarded cousin in what might be regarded as openly confrontational fashion. A different explanation might be, however, that, possibly, he had come to consider the idea of hard heredity as too distracting or controversial for a general public audience. Support for this conjecture seems provided by Galton's having taken the same course three years later when he made another appeal for public support, in the article "Hereditary Improvement," which appeared in the January 1873 number of *Fraser's Magazine*. Here despite his concurrently publishing scholarly papers promoting hard heredity, Galton took the tack of denying not only that he was identifying the hereditary castes he was advocating with currently extant social classes, and that his system would be anything but meritocratic, but also that he was advancing, and building upon, anything other than "the ordinary doctrines of heredity."[28] Seemingly the fact that Galton could so assiduously promote hard heredity before one audience and at the same time deny the propagation of such a doctrine before another might be perceived as expedient inconsistency, and more specifically as his manifesting a greater commitment to the success of his program than to the consistency of his science.

---

[28]"Hereditary Improvement," 124 for meritocratic sentiments, 116 and 123 for his assertion that he was building on the "ordinary doctrines of heredity."

5. *The nature of Galton's transfer to Pearson of authority for the doctrines and direction of eugenics.*

By the last decade of the nineteenth century Galton seems to have had two main purposes with respect to eugenics. The first was the further development and defense of his heredity tenets. Although he had discovered techniques, such as the coefficient of correlation, which would revolutionize all of science, he seems to have grown most fond of his "laws." Clearly they comforted him. To some extent this attachment seems to have waxed inversely with the waning prospects of eugenics' implementation. If, he seems to have felt, after all his efforts eugenics should end up on the library shelf, at least his Laws of Regression to the Mean and Ancestral Heredity and his doctrine of discontinuous evolution might endure.

His other great concern was finding a successor for the leadership of his eugenics effort, for Galton was by no means throwing in the towel there. Indeed, as Donald MacKenzie has shown, up to 1892 he made entreaties to at least seven promising young mathematicians, all Cambridge graduates, whom he appears to have been evaluating as prospective successors. Galton succeeded in interesting all in a particular problem related to his eugenics work, yet each, after having investigated his problem mathematically, returned to his own pursuits. Of these seven, the lack of interest of F. Y. Edgeworth, a distant cousin, must have proved the most frustrating, given his great mathematical ability, the proximity of his social views to Galton's, and yet his inability to share Galton's vision of eugenics as a promising means of social transformation.[29]

Thus Galton's sense of relief must have been substantial when, finally, he enlisted the support of Karl Pearson. Although thirty-five years Galton's junior, Pearson had already demonstrated a breadth of interests and depth of energy equaling even Galton's own. He had graduated from Cambridge with mathematical honors in 1879, passed the bar in 1881, secured an appointment as a professor of applied mathematics and mechanics at University College London

---

[29] MacKenzie, *Statistics in Britain*, 96-99.

in 1884 and a Gresham professorship in geometry in 1891. He had written (sometimes in German) on German history, philosophy and folklore. He was an idiosyncratic Socialist, a prominent feminist, and an almost compulsive controversialist. By 1893, at the age of 36, he had a hundred publications to his credit, nine of which were books. It is impossible to determine exactly when Pearson was perceived, and next enlisted, by Galton as his heir-apparent: Galton's modern-day biographers pass over the issue, and Pearson seems to have taken pains to make his and Galton's joining of forces appear to have occurred somewhat later (i.e., ca. 1900) than looks, from the nature of Pearson's publications, to have been the case.[30]

Whenever this momentous linkage happened, we must assume that it came with some provisos on Pearson's part, and one to have been his insistence on the right to make whatever revisions to Galton's doctrines he considered necessary to the success of eugenics. In 1893 Pearson had finished the editing, and extensively involved completion, of the second of two major statements by two recently-deceased physicists. No doubt he was considerably disin-

---

[30] For biographical considerations of Pearson, see Egon S. Pearson, "Karl Pearson: An Appreciation of Some Aspects of His Life and Work," *Biometrika* 28 (December 1936): 193-257, and "Part II: 1906-1936" of same, *Biometrika* 29 (February 1938): 161-248: M. Greenwood, "Pearson, Karl," *Dictionary of National Biography, 1931-1940*, ed. L.G. Wickham Legg (London: Oxford University Press, 1949), 681-84; Provine, *Origins*, 26-35; Churchill Eisenhart, "Pearson, Karl," *Dictionary of Scientific Biography*, vol. 10, Navashin-Piso, ed. Charles Coulston Gillispie (New York: Scribner's, 1974), 447-73; Bernard Semmell, "Karl Pearson: Socialist and Darwinist," *British Journal of Sociology* 9 (June 1958): 111-25; B.J. Norton, "Biometric Defense of Darwinism," 308-11; B.J. Norton, "Biology and Philosophy: The Methodological Foundations of Biometry," *Journal of the History of Biology* 8 (Spring 1975): 85-93; B. Norton, "Metaphysics and Population Genetics: Karl Pearson and the Background to Fisher's Multi-factorial Theory of Inheritance," *Annals of Science* 32 (1975): 537-53; Bernard J. Norton, "Karl Pearson and Statistics: The Social Origins of Scientific Innovation," *Social Studies of Science* 8 (February 1978): 3-34; Donald MacKenzie, "Statistical Theory and Social Interests: A Case Study," *Social Studies of Science* 8 (February 1978): 35-83; Donald MacKenzie, "Karl Pearson and the Professional Middle Class," *Annals of Science* 36 (March 1979): 125-43; Donald MacKenzie, *Statistics in Britain*, 73-93; G.R. Searles, "Eugenics and Class," in ed. Webster, *Biology, Medicine and Society*, 217-42; Kevles, *Name of Eugenics*, 20-40.

FIGURE 2. Karl Pearson, at ca. age 52, and Francis Galton, at age 87, at Fox Holm, Cobham. Reprinted from Karl Pearson, *The Life, Letters and Labours of Francis Galton*, vol. 3A (Cambridge: Cambridge University Press, 1930), Plate XXXVI, opposite p. 353.

clined to be constrained yet again by the dictates of someone else's system.[31] And apparently Galton acquiesced.

As much can be surmised from Galton's reaction when Pearson proceeded to rummage through his laws, unceremoniously throwing some out, next restoring some, and then reworking those which he had readmitted, often in radical fashion. Pearson first challenged the Ancestral Law in 1895, before he and Galton seem to have joined forces, but then, in 1898, claiming to be sufficiently impressed with the work Galton had done with basset pedigrees in 1897, he reversed himself, throwing out regression instead.[32]

---

[31]The two works which Pearson edited and completed were William Kingdon Clifford, *The Common Sense of the Exact Sciences* (London: Kegan Paul, Trench, 1885), and Isaac Todhunter, *A History of the Theory of Elasticity and of the Strength of Materials from Galilei to the Present Time*, 2 vols. in 3 (Cambridge: Cambridge University Press, 1886-1893).

[32]Pearson's first challenge to the Ancestral Law, which also included some consideration of regression, was his "Mathematical Contributions to the Theory of Evolution.-III. Regression, Heredity and Panmixia," *Philosophical Transactions* 187A (1896): 253-318 (received September 28, 1895; read November 28, 1895; revised November 29, 1895). See 254-98 for discussion of the Ancestral Law, 298-318 for regression. Pearson's most expansive reflection on the course of his considerations and revisions of Galton's laws to 1898 appears in his "On the Law of Ancestral Heredity," *Science*, n.s. 7 (11 March 1898): 337-39. A briefer development of the same report appears in his "Mathematical Contributions to the Theory of Evolution. On the Law of Ancestral Heredity," *Nature* 57 (10 March 1898): 452-53. The 1898 paper itself is Pearson, "Mathematical Contributions to the Theory of Evolution. On the Law of Ancestral Heredity," *Proceedings of the Royal Society* 62 (16 March 1898): 386-412 (received January 12, 1898, read January 27, 1898 with "A New Year's Greeting to Francis Galton, January 1, 1898"). For Pearson's jettisoning of regression, see 395-97. See also for Pearson's first uses of his newly redeveloped Ancestral Law, Cicely D. Fawcett and Karl Pearson, "Mathematical Contributions to the Theory of Evolution. On the Inheritance of the Cephalic Index," *Proceedings of the Royal Society* 62 (16 March 1898): 413-17.

For preliminary papers by Pearson laying out the statistical and conceptual groundwork for his statement of 1898, see Pearson, "Contributions to the Mathematical Theory of Evolution. —II. Skew Variations in Homogeneous Material," *Philosophical Transactions* 186A (1895): 343-414; Pearson and Alice Lee, "Mathematical Contributions to the Theory of Evolution. On Telegony in Man, &c.," *Proceedings of the Royal Society* 60 (9 December 1896): 273-83; Pearson, "Contributions to the Mathematical Theory of Evolution. Note on Reproductive Selection," *Proceedings of the Royal Society* 59 (18 June 1896): 301-05.

For Galton's earlier theoretical statements on regression, see Galton, "Typical Laws" (1878); his address as president of the anthropology section of the British Association, "Types and Their Inheritance," *Science* 6 (25 September 1885): 268-74. (The editors of *Science* based their text on "advance sheets of *Nature*," but no other publication seems to have been made of the address.) See more importantly, Galton, "Regression towards Mediocrity in Hereditary Stature," *Journal of the Anthropological Institute of Great Britain and Ireland* 15 (November 1885): 246-63, his classic memoir based upon the same data. See also Galton, "Family Likeness in Stature," *Proceedings of the Royal Society* 40 (21 January 1886): 42-63 (42-60 for regression, 60-63 for the Ancestral Law); "Hereditary Stature," *Nature* 33 (28 January 1886): 295-98, for extracts from Galton's presidential address to the Anthropological Institute, given on January 26, 1886; and Galton, "Hereditary Stature," *Nature* 33, letter to editor, (4 February 1886): 317, for his correction to these extracts; Galton, *Natural Inheritance* (1889), 95-119 and passim; Galton, "Discontinuity in Evolution," 364; see also "Questions Bearing on Specific Ability," *Nature* 51 (11 April 1895): 570-71, for discussion of a paper read by Galton at the Entomological Society on April 3, 1895.

For Galton's further development of the Ancestral Law, see Galton, "Family Likeness in Eye-colour," *Proceedings of the Royal Society* 40 (27 May 1886): 402-16; Galton, "Rate of Racial Change that Accompanies Different Degrees of Severity in Selection," letter to editor, *Nature* 55 (29 April 1897): 605-06; Galton's basset paper, "The Average Contribution of Each Several Ancestor to the Total Heritage of the Offspring," *Proceedings of the Royal Society* 61 (31 July 1897): 401-13; Galton, "Hereditary Colour in Horses," *Nature* 56 (21 October 1897): 598-99. See also Galton, "The Distribution of Prepotency," letter to editor, *Nature* 58 (14 July 1898): 246-47 for a further development of his theory of discontinuous evolution. See also for contemporaneous commentary on Galton's basset work, "A New Law of Heredity," 235-37, as in Footnote 4 of this chapter.

For a highly useful discussion of Pearson's operations on all of Galton's heredity doctrines, see Provine, *Origins*, passim, and especially "Appendix: Galton, Pearson and the Law of Ancestral Heredity," 179-87.

For coeval American critiques of Galton's rulings on regression and ancestral heredity, see W.K. Brooks, "Francis Galton on the Persistency of Type," 173-79; J. McKeen Cattell, review of Galton's "Average Contribution," 676-77; and Charles B. Davenport, "A History of the Development of the Quantitative Study of Variation," *Science*, n.s. 12 (7 December 1900): 864-70, which treats both Galton's and Pearson's contributions to that point. See also W.E. Castle, "The Laws of Heredity of Galton and Mendel, and Some Laws Governing Race Improvement by Selection," *Proceedings of the American Academy of Arts and Sciences* 39 (November 1903): 223-27; Castle, "On the Inheritance of Tricolor Coat in Guinea-Pigs and Its Relation to Galton's Law of Ancestral Heredity," *American Naturalist* 46 (July 1912): 437-40; and Edwin G. Conklin, "Phenomena of Inheritance," *Popular Science Monthly* 85 (October 1914): 322-25.

Moreover, he revised the Ancestral Law so extensively that although he insisted on binding Galton's name to the concept, it bore only a structural resemblance to the doctrine that Galton had officially introduced in 1885. Next in 1900 Pearson readmitted Galton's doctrine of regression but only after he had variously reworked it, giving it a moveable mean, thus permitting the claim that a stable eugenic upgrading, one not subject to the forces of regression to mediocrity, could be effected in as few as two or three generations.[33]

---

[33] By this point Pearson was concerned with revising the Ancestral Law so as to account for factors of selection, assortative mating and differential fertility. Moreover, he needed to develop better methods for the handling and correlation of qualitative traits. Regarding the latter concern, see Pearson, "Mathematical Contributions to the Theory of Evolution. – VII. On the Inheritance of Characters Not Quantitatively Measurable," *Philosophical Transactions* 195A (1901): 1-47 (received February 7, 1900; read March 1, 1900); Pearson and Alice Lee, "Mathematical Contributions to the Theory of Evolution. – VIII. On the Inheritance of Characters Not Capable of Exact Quantitative Measurement. Part I, Introductory; Part II, On the Inheritance of Coat Colour in Horses; Part III, On the Inheritance of Eye Colour in Man," *Philosophical Transactions* 195A (1901): 79-150 (received August 5, 1899; read November 16, 1899; withdrawn, rewritten and again received March 5, 1900). For the former concern, see Pearson, "Mathematical Contributions to the Theory of Evolution. On the Law of Reversion," *Proceedings of the Royal Society* 66 (22 March 1900): 140-64, with "A New Year's Greeting to Francis Galton.-January 1, 1900." (Received December 28, 1899; read January 25, 1900.) See also Pearson and Alice Lee, "Mathematical Contributions to the Theory of Evolution. On the Relative Variation and Correlation in Civilised and Uncivilised Races," *Proceedings of the Royal Society* 61 (5 July 1897): 343-57; Pearson and L.N.G. Filon, "Mathematical Contributions to the Theory of Evolution. – IV. On the Probable Errors of Frequency Constants and on the Influence of Random Selection on Variation and Correlation," *Philosophical Transactions* 191A (1898): 229-311; Pearson, Alice Lee, and Leslie Bramley-Moore, "Mathematical Contributions to the Theory of Evolution. – VI. Genetic (Reproductive) Selection: Inheritance of Fertility in Man, and of Fecundity in Thoroughbred Race-horses," *Philosophical Transactions* 192A (1899): 257-330; Mary Beeton and Karl Pearson, "Data for the Problem of Evolution in Man. II. A First Study of the Inheritance of Longevity and the Selective Death-rate in Man," *Proceedings of the Royal Society* 65 (7 October 1899): 290-305; Pearson, "Data for the Problem of Evolution in Man. IV. Note on the Effect of Fertility Depending on Homogamy," *Proceedings of the Royal Society* 66 (12 May 1900): 316-23; Mary Beeton, G.U. Yule, and Karl Pearson, "Data for the Problem of Evolution in Man. V. On the Connection between Duration of Life and Number of Offspring," *Proceedings of the Royal Society* 67 (31 October 1900): 159-79;

Pearson, Alice Lee, Ernest Warren, Agnes Fry, Cicely D. Fawcett and others, "Mathematical Contributions to the Theory of Evolution. – IX. On the Principle of Homotyposis and Its Relation to Heredity, to the Variability of the Individual and to That of the Race. Part I. Homotyposis in the Vegetable Kingdom," *Philosophical Transactions* 197A (1901): 285-379 (received October 6, 1900; read November 15, 1900).

Pearson's view of both regression and the Ancestral Law had grown more complicated by this point than the text indicates. By 1900 he had concluded that both worked well enough for those traits that he believed to be the product of blended inheritance, e.g., stature, but that the Ancestral Law could not apply to the study of those traits produced by exclusive inheritance, e.g., eye color, and that reversion would obtain in the place of regression in these cases. See Pearson, "Reversion," toto; Pearson and Lee, "Characters Not Capable of Exact Quantitative Measurement," pp. 98, 101-02, 120-21; Pearson, "Characters Not Quantitatively Measurable," passim; and Pearson, *The Grammar of Science*, 2d ed. (London: Adam and Charles Black, 1900): 478-80; 486-96. By 1903, however, after Pearson's biometric approach to the study of heredity had encountered impressive competition from Mendelism, these difficulties seem to have dissolved. Indeed, this enforced resolution of such major complexities and problems in the biometric approach could well have functioned as another demonstration of the precedence of eristic over scientific priorities for Pearson and, by virtue of association, Galton. See Pearson (with appendices by various authors), "The Law of Ancestral Heredity," *Biometrika* 2 (February 1903): 211-34; Pearson and Alice Lee, "On the Laws of Inheritance in Man," *Biometrika* 2 (November 1903): 357-462, especially 370. See also Pearson, "On a Criterion Which May Serve to Test Various Theories of Inheritance," *Proceedings of the Royal Society* 73 (7 May 1904): 262-80, passim.

Pearson's claim with respect to the ease of stable eugenic upgrading also experienced a steady evolution. Initially he contended that lines could be permanently remade in six to eight generations of careful breeding; see Pearson and Lee, "On the Laws of Inheritance in Man," 383, for a reference to this earlier view and the claim that his shift to the more optimistic one had been occasioned by the discovery that "parental correlation. . .is much higher than we could anticipate from Mr Galton's *Natural Inheritance* data." See also, however, ibid., 397, where the "extreme importance" of this shift is pointed out: "It emphasizes the all-important law that with judicious mating human stock is capable of rapid progress." Indeed: "A few generations suffice to modify a race of men, and the nations which breed freely only from their poorer stocks will not be dominant factors in civilisation by the end of the [twentieth] century." Moreover, see Bernard J. Norton, "Karl Pearson and Statistics," 28, for the observation that in reviewing *Natural Inheritance* before the Men's and Women's Club in 1889 Pearson objected that "the regression observed by Galton in the general population would not hold for long-selected lines." Possibly Pearson had seen W.K. Brooks's 1887 response to "Francis Galton on the Persistency of Type"; at any rate, it appears that Pearson had entertained a

This restatement, which served to make eugenics appear more immediately practicable, also enabled Pearson finally to throw out discontinuous evolution, the tenet that Galton had so strenuously defended eight years before. Pearson had long opposed this doctrine for personal and ideological reasons, as too analogous to political revolution and as a concept championed by archenemies (principally William Bateson) whom he had violently opposed for doing so.[34]

belief in the permanence of selected variations before embarking on his statistical investigations of biological matters. See Pearson, "On the Law of Ancestral Heredity," 402, for the claim that it would require six generations of directed breeding to reach "1.2 percent of truth ever afterwards." (At that point, in 1898, having discarded filial regression, Pearson was maintaining that a new and permanent ancestral center would be established after each generation of selection.) See also Pearson, *Grammar*, 481-86, for the intermediate claim that it would require four generations of selection for a line to breed true (within .9375 of a selected character) and to escape regression, which concept he had by then readmitted.

For Pearson's contention that no more than two to three generations of selection would be required to effect near-"truth" and escape regression, see Pearson and Lee, "On the Laws of Inheritance in Man," 382-83, 396-97; Pearson, "Law of Ancestral Heredity" (1903), 225-28; Pearson, "Prefatory Essay: The Function of Science in the Modern State," *Encyclopedia Britannica*, 10th ed., vol. 32; vol. 8 of new vols. (Edinburgh: Adam and Charles Black, 1902), vii-xxxvii. See ibid., xii for the claim that two to four generations would be required, but see also viii, where the standard given is "three, or at most four, generations. . . ." See also [Pearson,] "Francis Galton" (obituary), 443.

[34]For helpful discussions of Pearson's opposition to discontinuous evolution and/or the political impetus behind this position, see A.G. Cock, "William Bateson, Mendelism, and Biometry," *Journal of the History of Biology* 6 (Spring 1973): 10, 19-22; B.J. Norton, "The Biometric Defense of Darwinism," 283-88, 290-94; Ruth Cowan, "Nature and Nurture," 194, 196-97; Norton, "Karl Pearson and Statistics," 5-6, 11-12, 18-29; Donald MacKenzie, "Karl Pearson and the Professional Middle Class," 128-41, especially 134; MacKenzie and Barry Barnes, "Scientific Judgement: The Biometry-Mendelism Controversy," in eds. Barry Barnes and Steven Shapin, *Natural Order: Historical Studies of Scientific Culture* (Beverly Hills, CA: Sage Publications, 1979): 191-210 (toto); MacKenzie, "Sociobiologies in Competition: The Biometrician-Mendelian Debate," in ed. Webster, *Biology, Medicine and Society*, 266; MacKenzie, *Statistics in Britain*, 130-34, 146; Robert Olby, "The Dimensions of Scientific Controversy: The Biometric-Mendelian Debate," *British Journal for the History of Science* 22 (September 1989): 300-04, 319. See also, for an arresting demonstration of the psychogenic force and long standing of Pearson's abhorrence of revolutionary change, Pearson, "Anarchy," *Cambridge Review*, 30 March 1881,

Galton's response to all of this is the most noteworthy feature in the process. Through all of Pearson's jettisoning, readmitting, and drastic reworking he uttered virtually no protests or demurral, not publicly nor in letters that have survived. Certainly he failed to retract his tenets publicly, and privately he saluted Pearson for his revision of some, but, more important, by so silently permitting this reworking of some of his long-cherished tenets and dismissal of others, he engaged in one of the most remarkable and thorough-going capitulations in the history of science. Such could hardly be seen as the behavior of an independent and committed scientist. Rather Galton gave every appearance, and rightly so, of functioning as one far more concerned with the prosecution, survival, and success of a particular program.

(Indeed, in a letter to Pearson which the American hereditarians could hardly have known to exist—Pearson having chosen to leave it out of the Galton biography—Galton gave an unmistakable indication of his priorities in this regard. In December 1903 he celebrated the claim which concludes Pearson and Alice Lee's paper "On the Laws of Inheritance in Man"—that but "a few generations suffice to modify a race of men...." "I can hardly express my scale of satisfaction at the new ground so firmly won by you," Galton reported. Pearson's ruling constituted "a most hopeful contribution towards the hopes of improvement in the nature of Man," he exclaimed, only next to express the reservation "whether strictly true or no, in the face of the long experience of breeders—who require some 7 or 8 generations [and] still continue breeding out." Galton made it clear in several places that he both rejected Pearson's ringing claim and delighted in its utility to the promotion of eugenics.)[35]

6. *Galton's and Pearson's one-sided framing of their inquiry into, and consideration of, causative factors in development.*

268-70.

[35]Pearson and Lee, "Laws of Inheritance," 397; Galton to Pearson, 5 December 1903, Folder 245/18F, Galton Collection.

This attribute figured prominently in their best-known work. Indeed, Alphonse de Candolle discussed such a tendency as the central defect of *Hereditary Genius*, in one of the series of letters which he and Galton exchanged after his presentation to Galton of an advance copy of his *Histoire des Sciences*:

> You habitually highlight, as the principal cause, heredity. When you speak of other causes they are indicated accessorily, and without seeking to distinguish what holds particularly for them or for each one of them. At long intervals you mention other causes. Thus one can read many pages where you demonstrate the influence of heredity, before encountering one line, like at the top of page 88, on the *social influences*. The very title of your work implies the idea of only studying heredity, its laws and consequences, or else you would have written: *On the effect of heredity and other circumstances as to genius*. Surely you have rendered true service to science but your point of view has been essentially that of heredity.

With this observation de Candolle looks to have functioned as the most tangibly influential of Galton's reviewers, for Galton subsequently provided his forthcoming work, *English Men of Science*, with the subtitle *Their Nature and Nurture*. He also worked this dichotomy into the title of a paper prefatory to the book, the title of one of his papers on twins a year later, and the text of much other work besides. As suggested by Raymond Fancher in his masterful examination of the interaction of these two men, Galton "was not about to let himself be criticized again by de Candolle for overlooking environmental factors and emphasizing heredity exclusively in his titles."[36] Although such an adjustment should be

---

[36]The translation of de Candolle's critique was provided by Fancher in "De Candolle and Galton," 348. De Candolle's letter of January 2, 1873 appears untranslated and in its entirety in Pearson, *Life of Galton*, vol. 2, 136-39. See Fancher's remark in "De Candolle and Galton," 349. Fancher discusses the three of Galton's titles which made mention of "Nature and Nurture," 349 and 352n). Permission to quote Raymond E. Fancher's translation of an excerpt from a letter written by Alphonse de Candolle to Francis Galton, 2 January 1873, as published in Professor Fancher's "Alphonse de Candolle, Francis Galton, and the Early History of the Nature-Nurture Controversy," *Journal of the Behavioral Sciences* 19

seen as more cosmetic than fundamental in nature, at least it can be said of Galton that he was concerned with softening the appearance of eristic one-sidedness in his work.

Though this was hardly the case with Pearson, it might also be noted that initially such a focused quality was mostly implicit in his investigations. Apparently the unprecedentedly—and, for many, disturbingly—mathematical character of his incursions into the field of biology served to draw attention away from the basic fact that, by treating all manner of phenotypic characters as data for the statistical comparison of familial generations, thereby Pearson was treating, and regarding, all physical development as the product of heredity alone.[37] Moreover Pearson appears to have avoided discussing, and thus directing attention to, this effective result.

But then, in the opening years of the twentieth century, in making his only major contribution to the body of received "proofs" comprising the hereditarian argument, Pearson made this one-sidedness starkly explicit. In papers issued from 1901 through 1903, he reported the results of an investigation in which he had correlated both the physical and the mental characters of several thousand pairs of young siblings, a Middle-English word whose denotation he had modified for the project. (The physical characters were health, eye color, hair color, hair curliness, and cephalic index, head length, head breadth, and head height, as measured with specially designed "headspanners." The mental characters were vivacity, assertiveness, introspection, popularity, conscientiousness,

---

(October 1983): 341-52 (348), was kindly granted by John Wiley & Sons, Inc.

[37]Donald MacKenzie and Barry Barnes made much the same point in 1979, noting that with his statement of 1896 that "Given any organ in a parent and the same or any other organ in its offspring, the mathematical measure of heredity is the correlation of these organs for pairs of parents and offspring," Pearson was "thus building hereditarianism into the basis of the overall model of evolution." In this fashion, they write, "the study of phenotypic resemblances could proceed independently of any consideration of the processes by which these resemblances were produced." MacKenzie and Barnes, "Scientific Judgment," 194. Note as well that such an objection was directed toward Galton by a professor at Lake Forest College (then styled Lake Forest University) in Illinois. See Hiram M. Stanley, "Mr. Galton on Natural Inheritance," letter to editor, Nature 40 (31 October 1889): 642-43. Pearson is quoted from "Regression, Heredity, and Panmixia," 259.

temper, apparent intellectual ability, and handwriting. All of these characters were measured or rated by participating teachers in the children's schools. Pearson and his assistants then used these ratings and measurements to calculate the degree of resemblance in each trait shown by each pair of siblings.) Because these siblings' various physical and mental qualities had correlated, when averaged together in their respective physical and mental columns, to approximately the same degree (circa .5), Pearson found encouragement to insist that "we are forced to the perfectly definite conclusion: *That the mental characters in man are inherited in precisely the same manner as the physical.* Our mental and moral nature is, quite as much as our physical nature, the outcome of hereditary factors."[38]

---

[38]Pearson's major statements regarding this project were Pearson, "On the Inheritance of Mental Characters in Man," *Proceedings of the Royal Society* 69 (24 December 1901): 153-55; and Pearson, "On the Inheritance of the Mental and Moral Characters in Man, and Its Comparison with the Inheritance of the Physical Characters," *Journal of the Anthropological Institute of Great Britain and Ireland* 33 (July-December 1903): 179-237. The latter paper comprises the text of Pearson's Huxley Memorial Lecture for 1903, as delivered before the Anthropological Institute on October 16, 1903. The statement by Pearson comes from the 1901 paper, p. 155. A quite similar conclusion was expressed in the 1903 paper, on p. 204.

Because of varying completeness in the returns collected and a major redirection in traits to be considered, it is impossible to establish exactly how many sibling pairs were utilized for the final report. (Pearson nowhere supplies a number.) The best estimate of the number of pairs studied overall for the various traits considered is between 3,800 and 4,800. At the time of the 1901 report Pearson had parents measuring or rating their own children's height, length of forearm, arm span, and eye color, and teachers measuring or rating cephalic index, hair color and health. The mental characters in 1901, all rated at school, were intelligence, vivacity, conscientiousness, popularity, temper, self-consciousness, and shyness. After nearly one thousand sibling pairs had been studied, the first three physical traits mentioned were dropped from the study, to be replaced by the others specified in the text, and records from parents were no longer utilized.

In 1903 there was also one other physical character rated — athletic power (ability) — but Pearson refused to allow this character to be averaged with the other physical traits, obviously because its index of correlation was unsuitably high, at .72 for brother-brother pairs, .75 for sister-sister pairs, and .49 for brother-sister pairs. Including these results would have raised his overall physical-character correlation mean for these three sibling classes to .558, .553, and .508, respectively, which would have rounded off to .6 for the first two classes, and prevented him from making his kernel argument. See p. 199 for his explanation of his disqualification of this character.

Yet, as various critics remarked, it was only by wielding Occam's razor and the highly questionable proposition that like results demonstrate like causes that Pearson could advance the claim that such qualities as conscientiousness are as much the product of heredity, and as little the result of environment, as is eye color. Certainly American hereditarians could and did make much of Pearson's study, but, as will be shown, they were keenly alert to the aggressively, eristically, one-sided quality of such an operation as well.[39]

---

[39]For criticisms of Pearson's sibling study, see T.D.A. Cockerell, "The Inheritance of Mental Characters," letter to editor, *Nature* 65 (16 January 1902): 245; F.A.D., "Inheritance of Psychical and Physical Characters in Man," *Nature* 70 (9 June 1904): 137-38, where the author conjectures that "some, indeed, may incline to the opinion that he proves too much," and asks "if the influence of heredity is supreme alike in the mental and moral, and in the physical domain, what room is left for the action of teaching, training, discipline, and the environment generally, influences which the common experience of mankind has held to be of importance?" (138) For Pearson's response to critics in this period see Pearson, "The Inheritance of Mental Characters," letter to editor, *Nature* 65 (16 January 1902): 245-46, a response to Cockerell; and then Pearson, "The Inheritance of Mental Characters," letter to editor, *Nature* 65 (27 February 1902): 391, a response to a reply by Cockerell which includes this remark, addressed to Cockerell but obviously intended for all who would criticize the study and its conclusions: "Prof. Cockerell belongs to those critics who live in the region of 'may-be.' If he will collect observations on some 5000 to 6000 [sic] children as we have done, he may still come down from the region of 'may-be' and be able to place fact against fact."

Finally there was also the response of G. Archdall Reid, M. D., who noted in 1911 that Pearson "makes — not once but repeatedly, not only in scientific memoirs but also in popular lectures and letters to newspapers — the *unqualified* statement that the mental and physical characters of man are inherited at the same rate." Such a claim was meant to serve a purpose for Pearson, Reid claimed:

> It leads him to a false opposition between 'nature' and 'nurture,' instead of to the really obvious truth that the nature of man, the educable animal, is such that he is supremely responsive to nurture. It leads him to the notion that the poorer classes in England are, on the average, by nature inferior to their more fortunate compatriots, and thence to dire predictions concerning our future as a nation and to demands that something shall be done.

The problem with Pearson's argument, Reid contended, lay in its faulty

7. The nature of Pearson's social analyses as Galton's successor.

As the fifth feature, above, should suggest, clearly Karl Pearson would have been perceived as enunciating scientific doctrine for Galton, with Galton's permission and approval. Thus it seems not unreasonable to surmise that many onlookers could have assumed as well that various social views that Pearson was openly expressing were ones in which Galton concurred, and perhaps even ones which he had been quietly subserving all along.

Certainly Pearson was far less inclined than Galton to work by indirection. In 1902 he publicly exposed several of what could have been perceived as the deeper purposes of eugenics, for all to see. Writing a "Prefatory Essay" on "The Function of Science in the Modern State" for the tenth edition of the *Encyclopedia Britannica,* Pearson described "the special truth that dawned upon us at the end of 19th century" (a truth which, of course, had actually dawned four decades before): that "ultimately the training of even the apparently most insignificant workers in the community. . .[is] vital to a nation

---

employment of logic:

> The biometric plan of ascertaining correlations between variations, and thence surmising a causal connection. . .is merely a variant method of the very old method of concomitant variations which is described in almost every book on logic and...almost every work on the methods of science. There is, however, this difference: according to the method of correlated variations as exemplified by biometricians, if two things vary together *on the average*, there is *invariably* a causal connection between them; according to the method of concomitant variations as described by logicians, if two things *invariably* vary together, there is *probably* a causal connection.

See Reid, "The Inheritance of Mental Characters," letter to editor, *Nature* 88 (14 December 1911): 210-11. Moreover, for a much more extensive response, one accusing Pearson of drawing "perilously near the borders of Charlatanism," see Reid, "Methods of Research," *Eugenics Review* 3 (October 1911): 259-62. For agreement with Reid, at least on the question of Pearson's conclusions, see R. H. Lock, "Methods of Research, Being an Attempt to Reconcile the Views of Dr. G. Archdall Reid with Those of Other Biologists," *Eugenics Review* 3 (January 1911): 344-45.

in the evenly balanced contest of modern civilisation."[40] Indeed, he insisted, "the most intelligent nations will be victorious in [this] struggle; and it befits each state that would be great to-morrow as well as to-day to educate and organise itself, from the statesman at the top to the ploughboys at the basis." Yet Pearson went on to stress that although "it is desirable to have some preliminary classification of what work an individual is best suited for. . . it is, on the other hand, not only undesirable, but impossible, to subject every individual in the nation to a test of fitness for every possible calling." Pearson had a far simpler solution in mind: "With rough practical efficiency," he assured, "a man's work in life is settled by his caste or class." Besides, he warned, "it is cruel to the individual, it serves no social purpose, to drag a man of only moderate intellectual power from the hard-working to the brain-working group. . . ." After all, he advised, "we have to remember. . .that the middle class of England, which stands there for intellectual culture and brainwork, is the product of generations of selection from other classes and in-marriage."[41] Thus he insisted that the "first funda-mental principle in the education of a community" is that "the edu-cation must be specialized for each individual class of workers. . . ." "Class intelligence," he declared, ". . .must be at the basis of national fitness. . . .The nation needs that the great bulk of its members shall work at the same task as their forefathers. . . .The increase of the intellectual proletariat is a sign not of efficiency but of chaos in the national education." Perhaps not surprisingly, another part of the solution for Pearson was "the training of an oligarchic class in statecraft," which he described as "the first and perhaps hardest task of the modern state." Clearly Pearson could be read as implying that eugenics sanctioned, embodied, and perhaps to some extent had even been designed to foster a determined resistance to meritocratic mobility, at least from the lower classes upward.[42]

---

[40]Pearson, "Function of Science," ix.

[41]Ibid., ix, x, x, x, x.

[42]Ibid., x, x, x, x, xi, xii, xii. In the same vein, the English statistician and eugenicist R.A. Fisher advocated control of the membership in the learned professions such as law and medicine by professional societies, so that present members could keep

8. *The apparent insincerity of Galton's general call for a negative eugenics.*

From "Hereditary Talent and Character" forward Galton advocated, as the negative half of eugenics' social solution, the elimination of perceived inferior strains. In "Hereditary Improvement," the 1873 public appeal previously mentioned, he expounded a program in which a supposedly genetically superior and carefully endogamous population deliberately limits the existence of its more numerous and biologically inferior neighbors, eventually refusing them all rights to reproduce.[43] From that point forward, Galton called for such measures as one of the foremost purposes and goals of eugenics. Yet could he have genuinely desired that such a policy be implemented? Curiously, this is a question that is never raised in present- day discussions of Galton's eugenics, although it goes to the heart of his true purposes. In resolving such an issue we should ask,

---

out "new blood" from the wrong classes and conserve the professional positions for their own children. See Fisher, "Positive Eugenics," *Eugenics Review* 9 (February 1917): 206-12. Donald MacKenzie, who afforded a lead to this article in his *Statistics in Britain*, 184 and 279n., has expressed a similar view of the status-saving purposes of English eugenics: Eugenics, he writes, sanctioned "a rigidly stratified educational system. . .with only the narrowest of ladders to allow the unusually gifted child, the 'sport', to rise from the lower class." Thus, "eugenics offered the professional middle class an educational philosophy which enabled them to justify the effective monopoly of professional education by the existing professional class." See MacKenzie, "Eugenics in Britain," *Social Studies of Science* 6 (September 1976): 510.

Finally, note in this regard a brief article that Galton composed for American readers in 1885. His "Heredity in Business Merit" made a generally antimeritocratic appeal, arguing that in the selection of candidates for positions of responsibility ancestral tendencies should be considered alongside the results of competitive examinations. See *Journal of Heredity* 1 (October 1885): 4-6.

[43]Galton, "Hereditary Improvement," 123-30 for his description of the general scheme, 129 for his spelling out of the final solution: "I do not see why any insolence of class should prevent the gifted class, when they had the power, from treating their compatriots with all kindness, so long as they maintained celibacy," he writes. "But if these continued to procreate children, inferior in moral, intellectual and physical qualities, it is easy to believe the time may come when such persons would be considered as enemies to the State, and to have forfeited all claims to kindness."

as Darwin did of Galton upon reading the article, from where, then, the labor force would have come. Indeed: could Galton have seriously looked toward an England expunged of all its supposed biological inferiors, an England propelling itself towards economic suicide, or one in which his nephews' and nieces' progeny might, of necessity, be pressed into service as chimney sweeps and factory hands? [44] Clearly Galton must have been instrumentally insincere in advocating this major element of his eugenic solution.

Such a recognition dictates the phrasing of two further questions: Why then did Galton advocate such a policy, and what might the real goals and purposes of eugenics have been? It seems reasonable to assume that some of the more discerning (and cynical) British and American observers of Galton's own period could have speculated that Galton had devised and advocated so impracticable a policy as a ploy permitting a fallback position, as part of a deeper strategy which was fundamentally political in nature. Accordingly another scenario suggests itself, one not much more fantastic than Galton's published one. It must be stressed that no evidence exists to establish either that Galton necessarily conceived of such a train of events or that his American admirers perceived him as having done so. Rather, the following scenario represents simply a logical extrapolation from the tendencies thus far discernible in Galton's development of his heredity tenets and eugenics generally: If Galton or his successors could succeed in establishing the eugenic point of view — if they could convince the voting public (previous to the establishment of universal suffrage) that heredity determines intelligence and ability, that the existing social striations are

---

[44]For Darwin's question, see Darwin to Galton, 4 January 1873, in Pearson, *Life of Galton*, vol. 2, 176. Strictly speaking, Darwin was addressing the labor problems that Galton's new caste would encounter once set off from the larger community, but his concern would apply all the more if and when this faction succeeded in seizing control of society generally and began forcing its ostensibly biologically inferior numerical superiors out of existence.

Note that Henry Fawcett, operating entirely independently, of course, had already suggested one possible solution to this dilemma, should Galton really need one: that Britain form a permanently powerless laboring caste composed of Chinese coolies imported strictly for the purpose. See Fawcett, *The Economic Position of the British Labourer* (Cambridge and London: Macmillan and Co., 1865), 255-56.

biologically ordained, that an immutable inferiority inheres to the lower classes, and that the occasional superior sports manifesting themselves there constitute, if elevated, a racial hazard—at that point the logical corrective should suggest itself naturally, that these classes should be firmly restricted, and indeed, ideally, prevented from reproducing themselves. Next, however, equally naturally, after this last-mentioned notion had occasioned great hostility, undoubtedly someone would indicate that part of the plan's utter impracticability at any rate. Thereupon the proposition could be tendered to the lower orders that they may breed after all, and breed prolifically, and be fully educated to their proper station, and prosper as much as they are able, all as long as they and their progeny agree never to aspire to the professions or to any share in their own governance. Thus a perpetually inferior status, an eternal relegation to a subservient social and political caste, could be proffered to the lower orders as a relative favor, of sorts. Surely it should seem absurd to think that such a scenario could ever have actually been considered, or that such a bargain could be tendered, much less struck. But in so fundamentally altered a collective mindset as one which could be created by the lodgment, in the public mind, of the hereditarian dogma and its corollaries, such a concatenation of developments might be more conceivable.

9. *Problems of doctrinal control in Galton's final decade.*

The last ten years and few days of Galton's life, 1901 to January 17, 1911, comprised a period in which eugenics finally blossomed into a bustling social movement and was riven by abysmal conflict. Indeed, it was one of the closing ironies of his life that two hard-heredity-based eugenics movements struggled into existence, one in England, one in America, both claiming him as their founder (and accepting discontinuous evolution), yet both of which he became duty-bound to repudiate. Some commentators have implied that the major reason for this result was Karl Pearson, that in choosing so forceful a successor Galton had dictated the rupture that resulted between the anointed leadership of eugenics and its would-be faithful. To some extent, on the surface especially, such a view exhibits some elements of truth. Pearson was all-too-easily offended, and ferociously combative, as a number of American eugenicists had cause to note. Nevertheless, most of the blame for the schisms that resulted seems to have belonged to Galton himself, who in his final decade seems to have lost sight of some crucial considerations.

Perhaps Galton after ensuring that the scientific side of eugenics was in capable hands, had grown too anxious that before his demise, it would get underway as a popular movement as well. Certainly there are indications that he was growing concerned in this same period that Pearson's biometrical investigations were becoming increasingly isolated in the larger scientific community, as well as increasingly incomprehensible for the well-educated public.[45] For whatever combination of reasons, in October 1905 he took the highly significant step of forming an advisory committee for his Eugenics Record Office (whose operation was separate from Pearson's Biometric Laboratory, although both were housed at University College London).[46] The excuse he tendered to Pearson for the formation of this committee was that such a group might give Edgar Schuster, the chief researcher at the Office, incentive for a more energetic prosecution of his responsibilities.[47] What Galton seems

---

[45]See Forrest, *Francis Galton*, 277; Galton to Pearson, 8 March 1908, in Pearson, *Life of Galton*, vol. 3A, 334-35; and Galton to Pearson, 1 January 1908, ibid., 332-33.

[46]For essential information on the founding of the advisory committee, see Galton to Pearson, 30 October 1905, pp. 2-3, Folder 245/18F, in Galton Collection, explaining that the first meeting was to be held that evening at his home, and that the ten officers of this committee, which would have no regular chairman, would be Sir John Cockburn, F.W. Mott, W. Palin Elderton, V.V. Branford, Galton and Pearson's close friend and collaborator in eugenics research W.F.R. Weldon, a Dr. McDougal, a Dr. Muir (possibly David Mair), Galton himself, J.W. Slaughter, and C.W. Saleeby. Galton invited Pearson to the meeting, but not to serve on the committee. Perhaps this was for the reason indicated by Forrest in *Francis Galton*, 260—primarily that the Principal of the University of London, Sir Arthur Rücker, had warned Galton that if Pearson were permitted to interact with the Research Fellow doing the work of the Eugenics Research Office (Edgar Schuster) he would either "dominate. . .or quarrel with him."

[47]"They would help and encourage Schuster, [and] enable him all the better to get on with the 'Noteworthy Families' work during this winter. He is pleased with the

not to have realized sufficiently was that, in creating this committee, he was to some extent redistributing, and certainly dichotomizing, the authority he had given to Pearson for the development of eugenics doctrine. Indeed, several of the ten individuals who were appointed to this committee went on within two years to form a Eugenics Education Society.[48] Initially Galton extended an offer of assistance to the Society, then he abruptly reversed course upon hearing that sexual problems had been a topic of discussion at one of their meetings. In March 1908 Galton had cause to regret quite deeply his affiliation with the EES, when J. W. Slaughter, its chairman, was convicted of indecent assault. Slaughter's conviction was overturned on appeal, however, and soon thereafter Galton's safely respectable neighbor, the lawyer Montague Crackanthorpe, assumed the acting presidency of the EES and invited Galton to serve as its first honorary president. Galton acceded to the request, giving the Society all the more appearance, before the public, of authority over eugenics.[49]

---

idea [and] will act as *host*." Galton to Pearson, 30 October 1905, p. 1.

[48]These individuals were Cockburn, Mott, Slaughter and Saleeby. Slaughter was elected Chairman of the Provisional Council overseeing the formation of the Society, Cockburn chaired the first general meeting, and Saleeby was a prominent spokesman for the group. See Faith Schenk and A.S. Parkes, "The Activities of the Eugenics Society," *Eugenics Review* 60 (September 1968): 142-43. See Pearson, *Life of Galton*, vol. 3A, 339 for the text of a rough draft of a letter from Galton to Montague Crackanthorpe, dated 16 December 1906, discussing the formation of "some association of capable men who are really interested in Eugenics. . . ." Clearly Galton had meant to direct this group, as indicated by his suggestion that it be centered in the Eugenics Record Office. In a letter to Pearson dated May 18, 1908, Galton discussed the operations of the EES, which had started up on February 14,1908, with cautious optimism, explaining that "I have not *yet* ventured to join it, but as soon as I am assured that it is *safe* management, shall do so." Ibid., 339.

[49]See Pearson, *Life of Galton*, vol. 3B, 628, and Forrest, *Francis Galton*, 275, for details on Galton's misgivings. For the apparent source of Forrest's information on Slaughter, see Pearson to Galton, 26 March 1908, in Pearson, *Life of Galton*, vol. 3A, 335-36. Apparently Forrest saw the original of this letter in the Galton papers, for Slaughter's name is deleted in Pearson's text.

John Willis Slaughter, a native of Alabama, received an A.B. and B.D. from Lombard College in 1898; and a Ph.D. in philosophy (psychology) at Michigan, under the Scots emigré Robert Mark Wenley, in 1901; was an Honorary Fellow in 1901-02 and a docent in Aesthetics and the Philosophy of Evolution in 1902-03, at Clark; an instructor in psychology at the University of Cincinnati, 1903-04; and again a docent in Aesthetics at Clark, 1904-05. In the fall of 1905 he removed to England, where (owing in part perhaps to his possession of stenographic skills) he was named secretary of the Sociological Society. He also gave a series of extension lectures at the University of London on the subjects of educational sociology and the psychology of the youth and adolescence, in 1906-07, and other such courses in the same setting through 1912; was a lecturer for the Education Commission of the London County Council, a founding editor of the Eugenics Education Society's *Eugenics Review*, and founder of the (English) Moral Instruction League. After returning to the United States he gave lectures in South America and Mexico, 1912-15; was an outspoken opponent of American military involvement in Mexico in 1916; a general lecturer in the United States on various political subjects, 1915-19; editor of *The Public*, 1917-19; and a lecturer on civics and philanthropy at the Rice Institute, Houston, Texas, from that time forward. His one book, *The Adolescent* (1910) provides a treatment of G. S. Hall's subject matter with its soft heredity bases carefully muted.

Perhaps the most intriguing aspect of Slaughter in the present context has to do with his immediate prominence in English eugenics circles. Although Hall described him in 1910 as "an American who was taken over there to lead this work," in 1905 Slaughter attributed his remarkable ascent to the phenomenon whereby "an American seems to have the inside track, and does not suffer from the extreme social stratification that restricts the Englishman." The question remains as to how someone associated with Hall's soft heredity regimen could have been accorded such front-rank status by Galton and in the English eugenics community generally. See "Slaughter, John Willis," *Who's Who in America* 11 (1920-21), ed. Albert Nelson Marquis (Chicago: A. N. Marquis and Co.), 2606; and "The Evolution of Mankind as seen in the Child and the Race," a four-page broadside issued by the University of London; G. Stanley Hall to Rose Woodallen Chapman, 18 April 1910; and J. W. Slaughter to G. Stanley Hall, 31 October 1905; all in Folder 12, Box 10 (renumbered B1-2-8), "Hall-Sanford Correspondence with Faculty, Proter-Taber," G. Stanley Hall Papers, Clark University Archives, Clark University, Worcester, Massachusetts. Also, "Carranza Lauded in Cooper Union," *New York Times*, 29 June 1916, 4. Note also with respect to Slaughter's editorship of the *Eugenics Review* that no specification of the editors' identities appears in the journal until 1920, that most of the correspondence files dating from the early period have been destroyed (according to Kathleen Hodson, "The Eugenics Review 1909-1968," *Eugenics Review* 60 (September 1968): 162), and that the assignment above of an editorship to Slaughter is based on a statement to that effect to be found in Hall's letter to Mrs. Chapman.

For information on Crackanthope's and Galton's interaction, and on Galton agreeing to serve as the first Honorary President of the EES, see Forrest, *Francis*

Soon the EES was pulling in a variety of directions, with some of its members even taking the tack of publicly and vociferously assailing Pearson and his laboratory's work, claiming Galton's support in doing so, and intimating that Galton had grown disenchanted with Pearson's scientific tendencies. (One especial irritant for various officers of the EES had been a study published by Pearson's group which denied the effect of alcohol abuse on the development of offspring.) Accordingly Galton was compelled to spell out in the London *Times*, first, his approval of Pearson's alcohol-effects study; next, his unmitigated support of Pearson; and, finally, the possibility of his resigning the honorary presidency of the EES.[50] Thus, probably somewhat abashedly, Galton had been

---

Galton, 276-77. Note that James Crichton-Browne was the functioning president of the Society by this time.

For the text of Galton's address before the EES, given at Crackanthorpe's home upon accepting the position, see "Eugenics," 1-2, as in Footnote 1 of Chapter Three. For good commentary on this speech, which was given on 25 June 1908, see Pearson, *Life of Galton*, vol. 3A, 347-49. See as well, for useful information on the EES, its founding, and its conflict with Pearson and Galton, G.R. Searle, *Eugenics and Politics in Britain* (Leyden: Noordhoff International Publishing, 1976), 10-19; Donald MacKenzie, "Eugenics in Britain," 499-532; Schenk and Parkes, "Activities," 142-43; and Pauline M. H. Mazumdar, *Eugenics, Human Genetics and Human Failings: The Eugenics Society, Its Sources and Its Critics in Britain* (London: Routledge, 1992), 7-8, 28-30. Note that a fully informed account of the founding of the EES has yet to be written.

See also for a useful account of the activities and historical influence of the EES, Greta Jones, *Social Hygiene in Twentieth Century Britain* (London: Croom Helm, 1986), passim.

[50]For Galton's letters to the *Times*, see Galton, "Alcoholism and Offspring," letter to the editor, *The Times*, 3 June 1910, 6, and Galton, "The Eugenics Laboratory and the Eugenics Education Society," letter to the editor, *The Times*, 2 November 1910, 9. The latter publication included Galton's statement that "I wish to take this opportunity of saying that I wholly approve of the fairness and scientific thoroughness of the laboratory work under the direction of Professor Pearson." See Pearson, *Life of Galton*, vol. 3A, pp. 396-400, 404-09, and Forrest, Francis Galton, 282-84 for invaluable detail on the increasing (and quite public) conflict between Pearson and the EES, and, finally, between the EES and Galton as well. Note that Galton's letter to *The Times* of 2 November 1910, above, is reproduced in Pearson, *Life of Galton*, vol. 3A, 406-07.

made to recognize afresh the truth which he seems to have better understood in earlier days, that, as put by Karl Pearson in 1930, "a 'confession' [i.e., a declaration of adherence to specified articles of faith] is requisite to hold together the members of a sect, and that without this there will be just as many creeds taught as there are individual propagandists."[51]

A similar problem had been posed by the emphatic endorsement, by the EES and the new American eugenics votaries, of Mendelism. A number of historians of science have analyzed the gladiatorial battle waged by Galton and Pearson's biometrical school and the Mendelians in the early years of the twentieth century, and have identified a variety of intrinsic and extrinsic reasons for what many of them have described as an unnecessary conflict. Many of these interpretations have obvious merit and yet what the discussants seem to have overlooked, by not considering the possibility of basally-directing programmatic purposes in Galton's and Pearson's eugenics, is the attendant possibility that Galton and Pearson had no choice but to oppose Mendelism in inflexibly confrontational fashion.[52]   Certainly American hereditarian

---

[51]Ibid., 407.

[52]For a sampling of entries in this fairly colossal debate, see A.H. Sturtevant, *A History of Genetics* (New York: Harper and Row, 1965), *passim*; Cyril D. Darlington, *Genetics and Man* (New York: Schocken, 1969), *passim*; Farrall, *English Eugenics Movement*, 54-105; Provine, *Origins*, 25-89; Froggatt and Nevin, "The Mendelian-Ancestrian Controversy," 17-24; Froggat and Nevin, "Galton's 'Law of Ancestral Heredity,'" 10-21; Cock, "William Bateson, Mendelism and Biometry," toto; Norton, "Biometric Defense of Darwinism," toto; Robert de Marrais, "The Double-Edged Effect of Sir Francis Galton: A Search for the Motives in the Biometrician-Mendelian Debate," *Journal of the History of Biology* 7 (Spring 1974): 141-74; Norton, "Biology and Philosophy," toto; Lyndsay A. Farrall, "Controversy and Conflict in Science: A Case Study—The English Biometric School and Mendel's Laws," *Social Studies of Science* 5 (1975): 269-301; D. MacKenzie and S.B. Barnes, "Biometriker versus Mendelianer. Eine Kontroverse und ihre Erklarung," *Kölner Zeitschrift fur Soziologie und Sozialpsychologie*, Sonderheft 18 (1975), 165-96; B. Norton, "Metaphysics and Population Genetics," toto; Searle, *Eugenics and Politics*, 16-18; Lindley Darden, "William Bateson and the Promise of Mendelism," *Journal of the History of Biology* 10 (Spring 1977): 87-106; Norton, "Karl Pearson and Statistics," toto; MacKenzie, "Statistical Theory and Social Interests," toto; MacKenzie, "Karl Pearson and the Professional Middle Class," toto;   MacKenzie and Barnes,

psychologists, to judge from their own experience, may well have understood a highly significant functional difference distinguishing the two methodologies: Mendelism, which constituted a conspicuously superior technique for analyzing individual cases of hereditary transmission and predicting probable outcomes, obviously placed great emphasis on individual and exceptional cases. Biometrics, conversely, by studying massive numbers of cases in actuarial fashion, directed its emphasis where it most needed to be for Galton's and Pearson's perceptible purposes—on large groups generally, and well away from troublesome individuals and complicating exceptions. More importantly still, as Galton and Pearson could hardly afford to share the doctrinal reins of eugenics with a popular movement, they could hardly permit eugenics to be yoked with a dynamic new science which might surge off in any direction scientifically indicated.

American hereditarian psychologists seem to have comprehended all of this, and thus Galton's associated inability to endorse their brand of eugenics as well. They continued to hail him as their inspiration and general guide, however unilateral their esteem might have seemed. It may have been ironic indeed that Francis Galton was forced to abjure two popular movements which claimed him as their leader, but such an outcome was structurally dictated, and this his American heirs seem to have understood.

With respect to the nine foregoing features there are two general points which may need to be emphasized. Certainly there is no way of proving that Galton designed eugenics as an impediment to democracy and the upward mobility of the English working classes, nor even that eugenics had been informed by a political impetus of

---

"Scientific Judgment," toto; Nils Roll-Hansen, "The Controversy between Biometricians and Mendelians: A Test Case for the Sociology of Knowledge," *Social Science Information* 19 (1980): 501-17; Daniel J. Kevles, "Genetics in the United States and Great Britain 1890-1930: A Review with Speculations," in ed. Webster, *Biology, Medicine and Society*, 202-07; G.R. Searle, "Eugenics and Class," ibid., 217-23; MacKenzie, "Sociobiologies in Competition," toto; MacKenzie, *Statistics in Britain*, 73-152; Jan Sapp, "The Struggle for Authority in the Field of Heredity, 1900-1932: New Perspectives on the Rise of Genetics," *Journal of the History of Biology* 16 (Fall 1983): 311-34; Olby, "Dimensions of Scientific Controversy," toto.

any sort. Nonetheless, his development of eugenics, its constituent elements, and Galton's and Pearson's promotion and defense of the eugenics program did exhibit numerous attributes which both could be and were interpreted by informed, concerned onlookers of the period, both English and American, in exactly this way.

CHAPTER SIX: THE AMERICANS' VIEW AND LIMITED USES OF GALTON

FINALLY, THE SPECIAL VIEW which American hereditarian psychologists appear to have taken of Galton merits consideration. Certainly it should be stressed at the outset that they seem to have seen Galton as sincere in his claims that mankind could be much improved through artificial selection, and that heredity is much more important than environment to the ontogenetic development of intelligence. Moreover, as psychologists they appreciated Galton as a major contributor to the psychology of individual differences and to twentieth-century science generally. He as much as anyone was responsible for the quantification of biology, zoology and numerous other fields that revolutionized these studies in such short order. Indeed, it was perhaps the most consequential irony of his entire career that he who seems to have set out with such decidedly political motives should have succeeded in contributing so resoundingly to science. To a great extent this may say more about the sciences themselves at that point than about Galton, that they should have required someone with such insistent and clearly defined purposes to push them into applied (and applicable) forms. At any rate, Galton was this figure, and the prestige he won in mounting the pedestal resulting from such a contribution made him luminously useful to the forces of hereditarianism everywhere.

Yet it was Galton the longtime promulgator of eugenics who was most of interest, and use, to these psychologists intent on doing the same work in America. It would run beyond the compass of the present discussion to detail the variety of their eugenics-promoting activities that might indicate the extent of his influence, but the nature of the uses they made of Galton in this respect can be considered.

The first thing to be observed about these uses is that most had limitations, some more restricting than others. The first, already mentioned, stemmed from the polemical barricade dividing their two camps. Because these psychologists could hardly hope to interact directly with their guide after 1901, they were constrained to seek after him in his writings. Thus it becomes all the more important to assess the degree of limitation imposed by his relative accessibility in this form. How much of Galton's writings could these American psychologists be expected to have known about? After all, a number of Galton's more revelatory statements, and several of those cited here, had been published in popular, rather than scientific, journals. Nonetheless, all of the psychologists rallying under his standard look to have had good access to extensive listings of his writings. Robert Yerkes and, apparently, Edward L. Thorndike, along with numerous other Harvard students, had been introduced to Galton, and duly proselytized for eugenics, by Charles B. Davenport, their zoology professor (and later the leading organizer of eugenical research in America), at various of his well-remembered Francis Avenue soirées.[1] Indeed, Davenport preached eugenics, and Galton, all over the campus and beyond. As he wrote to Galton in 1897, he had celebrated the master's seventy-fifth birthday by parading his work before his class on the experimental and statistical study of phylogenesis: "my arms ached for days," he reported, "in consequence of having to transport your books and serials containing your briefer articles from the libraries." Two months later Davenport reported to Galton that "some of the students were so much interested in your work that an oportunity [sic] was given me to present the same matter before a larger audience at Cambridge." "I find," he added, "that merely reading the titles of your papers, as they appear, for instance, in the Roy.

---

[1]That Yerkes was introduced to Galton and eugenics by Davenport is shown by his own testimony on the matter. See Yerkes to Davenport, 7 July 1911, Charles B. Davenport Papers, American Philosophical Society, Philadelphia, Pennsylvania, hereafter cited as Davenport Papers. That Thorndike received such an indoctrination is deduced from the observation that he took a zoology course in his second year at Harvard, as provided by Geraldine Jonçich (Clifford) , *The Sane Positivist: A Biography of Edward L. Thorndike* (Middletown, CT: Wesleyan University Press, 1968), 96n., and from the assumption that Davenport was his instructor. Even if this assumption should prove incorrect, it seems likely that Thorndike would have been apprised of at least one of Davenport's various advertisements of eugenics on the campus and in the community generally.

FIGURE 3. Edward L. Thorndike, presumably near 1936, thus ca. age 62. Reprinted, with permission, from *A History of Psychology in Autobiography*, vol. 3, ed. Carl Murchison (Worcester, MA: Clark University Press, 1936), p. xvi.

FIGURE 4. Robert M. Yerkes, presumably near 1932, thus ca. age 56. Reprinted, with permission, from a *A History of Psychology in Autobiography*, vol. 2, ed. Carl Murchison (Worcester, MA: Clark University Press, 1932), p. xvii.

Soc's list awakens a very great interest in the author."[2]

There were several such inventories available. The Royal Society's *Catalogue of Scientific Papers* listed forty-two of Galton's publications (in three editions covering the years 1800 to 1883) at the time of Davenport's letter, and went on to add seventy-two more (covering 1884-1900) by 1916. Certainly these were papers published in scientific journals only, but resourceful researchers might be expected to have consulted *Poole's Index to Periodical Literature* as well, which was published in 1882 and revised in 1891, and which bore seventy Galton entries. And if these were listed only alphabetically by title, anyone consulting the area of heredity would have come across entries for "Hereditary Talent and Character," "Hereditary Genius: The Judges of England between 1660 and 1865," and "Hereditary Improvement." More significant still are Galton's own listings of his works. In 1883, in his widely-read *Inquiries into the Human Faculty and Its Development*, he appended citations for fourteen papers discussed in the book, including "Hereditary Improvement," if not "Hereditary Talent and Character." The second edition of the book, issued in 1907, provided the same listing. Then in his autobiography, published in 1908, Galton provided an extensive listing of "Books and Memoirs by the Author" which did include "Hereditary Talent and Character." What is more, he made direct reference to this manifesto and the particulars of its publication at two different places in the text, even specifying of the two-part article that these "two preliminary papers ...contain the germs of many of my subsequent memoirs, the contents of which went to the making of the following books: *Hereditary Genius*, 1869; *English Men of Science*, 1874; *Human Faculty*, 1883; *Natural Inheritance*, 1889; and to my quite recent writings on Eugenics." Properly motivated readers were all but conducted to Galton's maiden elaboration of eugenics.[3]

---

[2]Davenport to Galton, 20 March 1897 (misdated 1896); Davenport to Galton, 16 May 1897; Folder 235, Galton Collection.

[3]Royal Society of London, *Catalogue of Scientific Papers* (1800-1863), vol. 2 (London: George Edward Eyre and William Spottiswoode, 1868), 763; Royal Society of London, *Catalogue of Scientific Papers* (1864-1873), vol. 7 (London: C.J. Clay and Sons, 1877), 733; Royal Society of London, *Catalogue of Scientific Papers*, (1874-1883),

A perhaps more significant potential limitation on the American hereditarians' ability to be fully enlightened by Galton's and Pearson's promotions of eugenics might be their capacity to scrutinize these figures' evidence once in hand. Was this class of investigators truly able to distinguish actual weaknesses and the forced quality of Galton's and Pearson's case for the importance of heredity? Or, conversely, might they have been as impressed as they proclaimed themselves to be—and thereby incapable of profiting from the eristic example set before them? Certainly, such questions are difficult to resolve. Deeply partisan promoters and defenders of a cause are usually disinclined to acknowledge, publicly, inadequacies in the proof presented for their cause. Such candor, when it appears at all, seems to emerge in response to extraordinary circumstances or compelling incentives.

However remarkably, on at least two occasions such conditions did exist, and deeply-committed American hereditarians elaborated in detail the deficiencies that they found in Galton's and Pearson's evidence. The first of these was the biologist and apologist for eugenics, Frederick Adams Woods. Known to many as "the American Galton," Woods was accorded the appellation for publishing, from 1902 through 1913, an extended examination of various biographical materials, and expounding several sweeping biological generalizations, all for the purpose of establishing the absolute predominance of heredity over environment in effecting intelligence and morality. It was at the outset of one of the more important of these biographical exercises, *Mental and Moral Heredity in Royalty* (1906), that Woods revealed his actual opinion of Galton's

---

vol. 9 (London: C.J. Clay and Sons, 1891), 741-42; Royal Society of London, *Catalogue of Scientific Papers*, vol. 12, "Supplementary Volume," (London: C.J. Clay and Sons, 1902), 260; Royal Society of London, *Catalogue of Scientific Papers* (1884-1900), vol. 15 (Cambridge: Cambridge University Press, 1916), 197-98. William Frederick Poole and William I. Fletcher, *Poole's Index to Periodical Literature*, rev. ed., Part I-A-J, 1802-1881 (Boston: Houghton-Mifflin, 1891), 587 for the titles listed. Galton, *Inquiries into the Human Faculty and Its Development* (New York: Macmillan and Co., 1883; 2d ed. New York: E. P. Dutton Co., 1907), vii-ix for both editions; see also Galton, *Natural Inheritance*, "Appendix A," 219. Galton, *Memories*, "Appendix: Books and Memoirs by the Author," 325-31, 289 for the quoted matter, 310 for the other textual reference.

evidence for heredity in *Hereditary Genius*. In attempting to position his own book as superior to all the arguments for heredity preceding it, Woods detailed the problems inherent in Galton's attempts to establish true literary eminence; the incompleteness of Galton's listing of eminent literary figures without eminent relatives; Galton's general tendency to load his case unfairly, as shown in his selecting John Adams as his only American statesman, clearly for his distinguished progeny; a lack of system in the selection of his cases ("one might say that he showed a preference for those who had eminent relatives"); and Galton's glaringly inadequate consideration of the influence of family patronage. Such "objections cannot be raised against the evidence contained within these pages," Woods assured his many readers. Clearly some American hereditarians who had failed to detect the deficiencies in Galton's evidence for heredity should have been alerted to their existence, from 1906 on, by Woods's book.[4]

The other such critic of hereditarian proofs was E. L. Thorndike, America's earliest (hard) hereditarian psychologist. Thorndike first

---

[4]Woods, *Mental and Moral Heredity in Royalty: A Statistical Study in History and Psychology* (New York: Henry Holt and Co., 1906), 7, 8, 8, 8, 8, and 8 for the quoted matter.

For other American counters to Galton's argument in *Hereditary Genius*, see Charles H. Cooley, "Genius, Fame and the Comparison of Races," *Annals of the American Academy of Political and Social Science* 9 (May 1897): 317-58; John M. Robertson, "The Economics of Genius," *Forum* 25 (April 1898): 178-90; George R. Davies, "A Statistical Study of the Influence of the Environment," *Quarterly Journal of the University of North Dakota* 4 (April 1914): 212-36; Edwin Leavitt Clarke, *American Men of Letters: Their Nature and Nurture* (New York: Columbia University Press, 1916); Charles Elmer Holley, "The Relationship between Persistence in School and Home Conditions," in *Fifteenth Yearbook of the National Society for the Study of Education* (1916), Part II (Chicago: University of Chicago Press, 1916), 8-119; and, most notably, the psychologist James McKeen Cattell's series of studies of the influence of heredity and environment on the production of American men of science, issued from 1894 through 1933, and perhaps best represented by Cattell, "Statistics of American Psychologists," *American Journal of Psychology* 14 (April 1903): 310-28; Cattell, "A Statistical Study of American Men of Science. III. The Distribution of American Men of Science," *Science* n.s. 24 (7 December 1906): 732-42; and Cattell, "Families of American Men of Science," *Popular Science Monthly* 86 (May 1915): 504-15.

discussed Pearson's sibling study, one of the most formidable weapons in the hereditarian arsenal, in his 1903 text on educational psychology. Not only did he praise the investigation as "perhaps the most valuable research in educational psychology yet made," he especially credited Pearson's crucial use of logic. To doubt Pearson's interpretation of his findings would be "rash," he warned. "To believe that the fraternal resemblances in mental traits are due to environmental influences which work to such an extent as exactly to counterfeit in amount the force of inheritance," Thorndike declared, "is hardly possible to a critical mind." Yet in 1914, after one of Pearson's adjutants unleashed a savage attack on the American school of eugenics, Thorndike repaid the favor by considering Pearson's study in a far harsher light. In the third volume of his revised educational psychology text, Thorndike pointed up, among a host of other problems, Pearson's inflexibility in interpreting his results:

> . . .unless one is a blind devotee to the irrespressibility and unmodifiability of original mental nature, one cannot be contented with the hypothesis that a boy's conscientious-ness or self-consciousness is absolutely uninfluenced by the family training given to him. . . .[T]o prove that conscientiousness is irrespective of training is to prove too much. One fears that Professor Pearson may next produce coefficients of correlation to show that the political party a man joins, the place where he lives, and the dialect he speaks are matters of pure inheritance. . . .

Wholly repudiating Pearson's study, Thorndike concluded, "The reader may at this stage be in some doubt as to precisely what Professor Pearson's measurements give as a probable similarity of brothers' original natures. I share this doubt."[5]

The final limitation concerns the relative irresistibility of Galton's influence. To judge by the record of affected psychologists' activities, Galton seems to have converted next to none at first blush,

---

[5]Thorndike, *Educational Psychology* (1903), 54, 54, 55; Thorndike, *Educational Psychology*, vol. 3 (1914), 242, 242.

nor to have moved any to take up the flag. Thorndike seems to have witnessed Davenport's evangelizing in 1896 or 1897, yet in 1900 he published a first book, *The Human Nature Club*, which showed him to have been at that point quite evenhanded in his treatment of the nature-nurture issue, if not a bit of an environmentalist. It was not until 1903 that he would emerge as an emphatic and insistent champion for eugenics.[6] Yerkes, whom Davenport exposed to

---

[6]Thorndike, *The Human Nature Club: An Introduction to the Study of Mental Life* (New York: Chatauqua Press, 1900). In the personae of several fictionalized spokesmen he argued against the heritability of criminality: The Jukes family was possibly produced by a poor heredity, "probably" by a poor antenatal nutrition, and "certainly" by a poor environment (192-93); criminals are no less intelligent than the mass of men (195); and only two percent of the inmates of English industrial (reformatory) schools are descended from habitual criminals (193). Moreover, morality is probably environmentally determined, and overall mental development is slightly more influenced by factors of nurture than by those of nature (186).

For evidence of Thorndike's dramatic swerve into hereditarian psychology and eugenics, see *Educational Psychology* (1903), 40-79, and the heavy-handed assistance he gave to students publishing papers in *Heredity, Correlation and Sex Differences in School Abilities*, Columbia University Contributions to Philosophy, Psychology and Education 11, Whole No. 2 (February 1903), ed. Edward L. Thorndike.

Regarding the course of Thorndike's rapid evolution into a committed hereditarian, note Thorndike's first and possibly only letter to Galton, written in late 1901 (the only Thorndike letter in the Galton Collection). Here Thorndike informed the master that he had recently conducted a study adducing evidence to indicate that the gifted were not marrying less, later, or with less issue than the rest of humanity; told him of Clark Wissler's study showing a low correlation between results on psychophysical mental tests and college grades; appealed for another printing of *Hereditary Genius*; and mentioned that "I myself am now giving various simple and convenient tests to well-born children and to children in reformatories in the hope of thereby finding tests which are significant of efficiency in life." There are several things to be said for this communication. First, note that the first investigation which Thorndike referred to was, as published in his "Marriage Among Eminent Men" (*Popular Science Monthly* 61 (August 1902): 328-29), a simple tabulation of marital and childbearing data from *Who's Who*; that his resulting article made only a vaguely eugenical-sounding statement, in its opening; and that its general thrust would seem contrary to the fear-mongering common to, e.g., Pearson's forthcoming publications on this same score. Moreover, even taking into account Thorndike's characteristically laconic style, the letter seems to lack any overt celebration of eugenics of the sort which Galton was accustomed to receiving from American enthusiasts. Yet Thorndike's description of his testing project and

Galton and eugenics in 1899, waited ten years before taking any action. In 1911 he wrote to Davenport reporting that for the past couple of years he had been capitalizing on the "opportunity to bring the facts [of eugenics] forcibly to the attention" of his students, and recently had been devoting one to two weeks to such promotion in three of his classes. More importantly, he urged the formation of a more aggressive eugenics propaganda effort, beginning with a new national organization associated more with physicians than farmers (an obvious reference to the American Breeders' Association). If Davenport could not attempt this work, he politely threatened, busy as he was, he might organize such a body himself. Clearly Yerkes was fully switched on to eugenics; clearly, as he testified to Davenport, it had been his introduction to Galton that had caused him to be "keenly interested in psychic heredity and its practical importance" ever since; yet equally clearly it had required something more than mere knowledge of Galton and his work to impel him to action.[7] The only eugenics-promoting psychologist who appears to have taken to eugenics upon first exposure was Lewis Terman, who had come across Galton in the course of his studies at Indiana in 1902, and even he did nothing to advance the cause until 1912, ten years farther on.[8]

---

statement of its purpose do suggest an emerging Galtonism in him. Conceivably Thorndike was seeking some general shibboleths, not so much for distinguishing "well-born" and less-advantaged subjects (that kind of sorting could be conducted virtually on the basis of apperances), but as dependable differences which could be presented as indices of hereditarily-determined intelligence and thus of inborn, immutable, worth. Apparently Thorndike was studying Galton closely now, and possibly with similar sociopolitical ends in view. See Thorndike to Galton, 4 December 1901, Folder 327/1, Galton Collection.

[7]Yerkes to Davenport, 17 July 1911, for the two quotations, 16 September 1911 for the urgings and polite threat, Davenport Papers.

[8]Terman, "Lewis M. Terman," 307, 311; Henry L. Minton, *Lewis M. Terman: Pioneer in Psychological Testing* (New York: New York University Press, 1988), 70.
     For one of the earliest published demonstrations of Terman's subscription to hereditarian psychology and eugenics, see Terman, "The Significance of Intelligence Tests for Mental Hygiene," *Journal of Psycho-Asthenics* 18 (March 1914): 119-27, especially 120. (Note that this paper was also published under the same title in *Transactions, International Congress on School Hygiene* 3 (1914): 502-08, having been

FIGURE 5. Lewis M. Terman, presumably near 1932, thus ca. age 55. Reprinted, with permission, from *A History of Psychology in Autobiography*, vol. 3, ed. Carl Murchison (Worcester, MA: Clark University Press, 1932), p. xvi.

given at that historic Buffalo, New York, convocation in August 1913.) The posing of 1912 as the year in which Terman's services to eugenics commenced is based on the present author's analysis of testing procedures Terman was implementing in California at this time, toward his revision of the Binet-Simon. The judgment that Terman was converted to hereditarianism upon exposure to Galton is based on the conclusion of his 1905 dissertation, a discussion of his testing of the intelligence of a number of school children, where Terman offered the largely gratuitous observation that, "while offering little positive data on the subject," his study had "strengthened" his "impression of the relatively greater importance of *endowment* over *training*, as a determinant of an individual's intellectual rank among his fellows." Terman, "Genius and Stupidity: A Study of Some of the Intellectual Processes of Seven 'Bright' and Seven 'Stupid' Boys," as published in *Pedagogical Seminary* 13 (September 1906): 372.

Given all these factors, the most important—and unrestricted—use these psychologists did make of Galton was as an exemplar. And here, again, they were uniquely equipped to see beneath the surface. Owing to their special apprehension of his deeper concerns and the fields in which he worked, this particular body of American admirers was able to appreciate the renowned inventor of eugenics as secretly operating eristically, as surreptitiously subordinating his declared goal, the perfecting of mankind, to the justification and establishment of an unimpeachable oligarchy, through the development and application of a modern-day Noble Lie. Galton demonstrated how properly-equipped covert politicians could cloak themselves in scientists' garb and pursue primarily political ends in ways not possible for declared politicians, by appropriating the authority (and general impenetrability) of modern science.

The final way Galton inspired American hereditarian psychologists is once more ironic. After all he had given them technically and shown them behaviorally, the contribution that seems to have charged them most vitally was his ultimate failure to carry the day. By failing to corroborate his crucial ontogenetic claim Francis Galton set out a veritable holy grail for his keenly ambitious heirs to pursue.

BIBLIOGRAPHY

Books, Monographs, and Dissertations

Albert, Prince Consort. *Addresses Delivered on Different Public Occasions by His Royal Highness The Prince Albert, President of the Society for the Encouragement of Arts, Manufactures, and Commerce.* London: Bell and Daldy, 1857.

_____. *Letters of the Prince Consort 1831-1861,* ed. Kurt Jagow. London: John Murray, 1938.

Andersson, C.J. *Lake Ngami.* London: Hurst & Blachett, 1856.

Aristotle. *Aristotle on Fallacies Or the Sophistici Elenchi,* trans. Edward Poste. London: Macmillan and Co., 1866.

Bagehot, Walter. *Parliamentary Reform-an Essay Reprinted, with Considerable Additions, from the National Review.* London: 1859.

Barker, Ernest. *The Political Thought of Plato and Aristotle.* New York: G.P. Putnam's Sons, 1906.

_____. *Political Thought in England: From Herbert Spencer to the Present Day.* New York: Henry Holt and Co., 1915.

_____. *Greek Political Theory: Plato and His Predecessors.* London: Methuen & Co., 1918.

Bateson, William. *Materials for the Study of Variation, Treated with Especial Regard to Discontinuity in the Origin of Species.* London: Macmillan and Co., 1894.

Baxter, R. Dudley. *National Income – The United Kingdom.* London: Macmillan and Co., 1868.

Best, Geoffrey. *Mid-Victorian Britain, 1851-1875.* New York: Schocken Books, 1972.

Blacher, L.I. *The Problem of the Inheritance of Acquired Characters: A History of A Priori and Empirical Methods Used to Find a Solution,* trans. Katy McKinin and Noel Hess; English translation ed. Frederick B. Churchill. New Delhi: Amerind Publishing Co., 1982.

Blacker, C.P. *Eugenics: Galton and After.* Cambridge, Mass.: Harvard University Press, 1952.

Briggs, Asa. *The Age of Improvement, 1783-1867.* New York: David McKay Co., 1959.

Broad, William, and Nicholas Wade. *Betrayers of the Truth*. New York: Simon and Schuster, 1982.

Carlyle, Thomas. *Shooting Niagara and After*. London: Chapman and Hall, 1867.

Clarke, Edwin Leavitt. *American Men of Letters: Their Nature and Nurture*. New York: Columbia University Press, 1916.

(Clifford), Geraldine Jonçich. *The Sane Positivist: A Biography of Edward L. Thorndike*. Middletown, CT: Wesleyan University Press, 1968.

Clifford, William Kingdon. *The Common Sense of the Exact Sciences*, ed. Karl Pearson. London: Kegan Paul, Trench, 1885.

Cowan, Ruth Leah Schwartz. "Sir Francis Galton and the Study of Heredity in the Nineteenth Century." Ph.D. diss., Johns Hopkins University, 1969.

Darlington, Cyril D. *Genetics and Man*. New York: Schocken, 1969.

Darwin, Charles. *The Variation of Animals and Plants Under Domestication*. 2 vols. London: John Murray, 1868.

_____. *The Descent of Man and Selection in Relation to Sex*. 2 vols. New York: D. Appleton & Co., 1871.

Darwin, Francis, ed. *The Life and Letters of Charles Darwin*. 3 vols. London: John Murray, 1887.

Davidoff, Leonore, and Catherine Hall. *Family Fortunes: Men and Women of the English Middle Class, 1780-1850*. Chicago: University of Chicago Press, 1987.

de Candolle, Alphonse. *Histoire des Sciences et des Savants depuis Deux Siècles suivie D'Autres Études sur des Sujets Scientifiques en Particulier sur la Sélection dans L'Espèce Humaine*. Geneva: H. Georg, Libraire-Éditeur, 1873.

Disraeli, Benjamin. *Sybil, or the Two Nations*. 3 vols. London: H. Colburn, 1845.

Eliot, George. *Felix Holt, The Radical*. Edinburgh and London: W. Blackwood and Sons, 1866.

Evans, J.D.G. *Aristotle's Concept of Dialectic*. Cambridge: Cambridge University Press, 1977.

Eyler, John M. *Victorian Social Medicine: The Ideas and Methods of William*

Farr. Baltimore: Johns Hopkins University Press, 1979.

Fancher, Raymond E. *The Intelligence Men: Makers of the IQ Controversy.* New York: Norton, 1985.

Farrall, Lyndsay. *The Origins and Growth of the English Eugenics Movement, 1865-1925.* New York: Garland, 1985.

Fawcett, Henry. *Mr. Hare's Reform Bill Simplified and Explained.* Westminster: Printed by T. Brettnell, 1860.

_____. *The Economic Position of the British Labourer.* Cambridge and London: Macmillan and Co., 1865.

Forrest, D.W. *Francis Galton: The Life and Work of a Victorian Genius.* London: Paul Elek, 1974.

Fox, Celina, Richard Johnson, Roy MacLeod, Edward Miller, and Gillian Sutherland. *Education,* ed. Gillian Sutherland. Dublin: Irish University Press, 1977.

Galton, Francis. *Tropical South Africa.* London: John Murray, 1853.

_____. *Hereditary Genius: An Inquiry into Its Laws and Consequences.* London: Macmillan and Co., 1869.

_____. *English Men of Science: Their Nature and Nurture.* London: Macmillan and Co., 1874.

_____. *Inquiries into the Human Faculty and Its Development.* New York: Macmillan and Co., 1883; 2d ed. New York: E.P. Dutton Co., 1907.

_____. *Natural Inheritance.* London: Macmillan and Co., 1889.

_____. *Hereditary Genius: An Inquiry into Its Laws and Consequences.* 2d ed. London: Macmillan and Co., 1892.

_____. *Memories of My Life.* London: Methuen & Co., 1908.

Gaskell, Elizabeth. *Mary Barton.* 2 vols. London: Chapman and Hall, 1848.

Halliday, R. J. *John Stuart Mill.* London: Allen and Unwin, 1976.

Hare, Thomas. *A Treatise on the Election of Representatives, Parliamentary and Municipal.* London: Longman, Brown, Green, Longmans, & Roberts, 1859.

Hessler, John Gerhard. "Victorians and the Threat of Democracy." Ph.D.

diss., Stanford University, 1977.

Hilts, Victor L. *A Guide to Francis Galton's English Men of Science*. Transactions of the American Philosophical Society n.s. 65, Part 5 (1975): 1-85.

_____. *Statist and Statistician*. New York: Arno Press, 1981.

Huxley, Leonard. *Life and Letters of Thomas Henry Huxley*. 2 vols. London: Macmillan and Co., 1900.

Huxley, Thomas. *On the Origin of Species: Or, the Causes of the Phenomena of Organic Nature. A Course of Six Lectures to Working Men*. New York: D. Appleton & Co., 1863.

Jones, Donald K. *The Making of the Educational System, 1851-81*. London: Routledge & Kegan Paul, 1977.

Jones, Greta. *Social Hygiene in Twentieth Century Britain*. London: Croom Helm, 1986.

Kay, Joseph. *The Social Condition and Education of the People in England and Europe; Shewing the Results of the Primary Schools, and of the Division of Landed Property, in Foreign Countries*. 2 vols. London: Longman, Brown, Green and Longmans, 1850.

Kevles, Daniel J. *In the Name of Eugenics: Genetics and the Uses of Human Heredity*. Berkeley: University of California Press, 1985.

Kingsley, Rev. Charles. *Alton Locke, Tailor and Poet: An Autobiography*. London: Chapman and Hall, 1850.

Lorimer, James. *Political Progress Not Necessarily Democratic; Or, Relative Equality the True Foundation of Liberty*. London and Edinburgh: Williams and Hargate, 1857.

MacKenzie, Donald. *Statistics in Britain, 1865-1930: The Social Construction of Scientific Knowledge*. Edinburgh: Edinburgh University Press, 1981.

Martin, Theodore. *The Life of the Royal Highness the Prince Consort*. 5 vols. 2d. ed. London: Smith, Elder & Co., 1875-79.

Mayr, Ernst. *The Growth of Biological Thought: Diversity, Evolution, and Inheritance*. Cambridge, Mass.: Belknap Press, 1982.

Mazumdar, Pauline M.H. *Eugenics, Human Genetics and Human Failings: The Eugenics Society, Its Sources and Its Critics in Britain*. London: Routledge, 1992.

Men of the Time: A Biographical Dictionary of Eminent Living Characters of Both Sexes. 6th ed. London: George Routledge and Sons, 1865.

Mill, John Stuart. *Thoughts on Parliamentary Reform.* London: Parker and Son, 1859.

_____. *Considerations on Representative Government.* London: Parker, Son and Bourn, 1861.

_____. *The Later Letters of John Stuart Mill 1849–1873*, ed. Francis E. Mineka and Dwight N. Lindley. Toronto: University of Toronto Press, 1972.

_____. *Essays on Politics and Society*, ed. J.M. Robson. Vols. 1 and 2. Toronto: University of Toronto Press, 1977.

Minton, Henry L. *Lewis M. Terman: Pioneer in Psychological Testing.* New York: New York University Press, 1988.

Musgrave, P.W. *Technical Change: The Labour Force and Education; A Study of the British and German Iron and Steel Industries 1860-1964.* Oxford: Pergamon Press, 1967.

Nettleship, Richard Lewis. *Lectures on the Republic of Plato.* 2d. ed., ed. Lord Charnwood. London: Macmillan and Co., 1901.

Norris, Rev. Canon. *The Education of the People.* London: Macmillan and Co., 1869.

Noyes, John Humphrey. *Essay on Scientific Propagation.* Oneida, NY: published by the Oneida Community, 1870(?).

Olby, Robert C. *Origins of Mendelism.* New York: Schocken Books, 1966.

Pearson, Karl. *The Grammar of Science.* 2d ed. London: Adam and Charles Black, 1900.

_____. *The Life, Letters and Labours of Francis Galton.* 3 vols. in 4. Cambridge University Press, 1914, 1924, 1930.

_____. *Francis Galton 1822-1922: A Centenary Appreciation.* London: Cambridge University Press, 1922.

_____. *The History of Statistics in the 17th and 18th Centuries Against the Changing Background of Intellectual, Scientific and Religious Thought: Lectures by Karl Pearson Given at University College London during the Academic Sessions 1921-1933*, ed. E.S. Pearson. London: Charles Griffin & Co., 1978.

Perkin, Harold. *The Origins of Modern English Society: 1780-1880*. London: Routledge & Kegan Paul, 1969.

Plato. *The Republic of Plato Translated into English with an Analysis and Notes*. Trans. John Llewelyn Davis and David James Vaughan. London: Macmillan and Co., 1891.

Popenoe, Paul, and Roswell Hill Johnson. *Applied Eugenics*. New York: Macmillan, 1918.

Popper, Karl R. *The Open Society and Its Enemies*. 2d ed. 2 vols. London: Routledge and Kegan Paul, 1952.

Porter, Theodore M. *The Rise of Statistical Thinking, 1820-1900*. Princeton: Princeton University Press, 1986.

Pound, Reginald. *Albert: A Biography of the Prince Consort*. London: Michael Joseph, 1973.

Provine, William B. *The Origins of Theoretical Population Genetics*. Chicago: University of Chicago Press, 1971.

Roper, Jon. *Democracy and Its Critics: Anglo-American Democratic Thought in the Nineteenth Century*. London: Unwin Hyman, 1989.

Russell, J. Scott. *Systematic Technical Education for the English People*. London: Bradbury, Evans, & Co., 1869.

Searle, G.R. *Eugenics and Politics in Britain*. Leyden: Noordhoff International Publishing, 1976.

Solomon, Barbara Miller. *Ancestors and Immigrants: A Changing New England Tradition*. Cambridge, Mass.: Harvard University Press, 1956.

Stoddard, Lothrop. *The Revolt Against Civilization: The Menace of the Under Man*. New York: Scribner and Sons, 1922.

Sturtevant, A.H. *A History of Genetics*. New York: Harper and Row, 1965.

Tankard, James W., Jr. *The Statistical Pioneers*. Cambridge, Mass.: Schenkman Publishers Inc., 1984.

Thomson, J. Arthur. *Heredity*. London: John Murray, 1908.

Thorndike, Edward L. *The Human Nature Club: An Introduction to the Study of Mental Life*. New York: Chatauqua Press, 1900.

_____. *Educational Psychology*. New York: Science Press, 1903.

_____, ed. *Heredity, Correlation and Sex Differences in School Abilities,* Columbia University Contributions to Philosophy, Psychology and Education 11. Whole No. 2 (February 1903).

_____. *Educational Psychology.* Vol. 3, Mental Work and Fatigue, and Individual Differences and Their Causes. New York: Teachers College, Columbia University, 1914.

Todhunter, Isaac. *A History of the Theory of Elasticity and of the Strength of Materials from Galilei to the Present Time,* ed. Karl Pearson. 2 vols. in 3. Cambridge: Cambridge University Press, 1886-1893.

White, Charles. *An Account of the Regular Gradation in Men, and in Different Animals and Vegetables, and from the Former to the Latter.* London: Printed for C. Dilly, In the Poultry, 1799.

Wiggam, Albert Edward. *The Fruit of the Family Tree.* Garden City, NY: Garden City Publishing Co., 1924.

Wollaston, William. *The Religion of Nature Delineated.* London: Sam. Palmer, 1724.

Woods, Frederick Adams. *Mental and Moral Heredity in Royalty: A Statistical Study in History and Psychology.* New York: Henry Holt and Co., 1906.

Wright, D. G. *Democracy and Reform, 1815-1885.* London: Longman Group, 1970.

ARTICLES, CHAPTERS IN WORKS BY VARIOUS AUTHORS, SYMPOSIUM PAPERS, SPEECHES, COURSEWORK ESSAYS, NEWSPAPER ARTICLES, AND LETTERS TO EDITORS

Alvarez Peláez, Raquel. "Las Fuentes Francesas de La Eugenesia de Galton." *Asclepio* 37 (1985): 165-81.

Annan, N.G. "The Intellectual Aristocracy." In *Studies in Social History: A Tribute to G.M. Trevelyan,* ed. J.H. Plumb, 241-87. London: Longmans, Green and Co., 1955.

Armytage, W.H.G. "W.E. Forster and the Liberal Reformers." In *Pioneers of English Education: A Course of Lectures Given at King's College, London,* ed. A.V. Judges, 207-26. London: Faber and Faber, 1952.

"Art. I." *Edinburgh Review*, July 1861, 1-20.

Ashburton, Lord. "At the Banquet in the Birmingham Town Hall, On the Occasion of Laying the First Stone of the Birmingham and Midland Institute." Address Given on 22 November 1855. In *Addresses Delivered on Different Public Occasions by His Royal Highness The Prince Albert, President of the Society for the Encouragement of Arts, Manufactures, and Commerce*, by Albert, Prince Consort, 170-73. London: Bell and Daldy, 1857.

Aydelotte, William O. "Patterns of National Development: Introduction." In *1859: Entering an Age of Crisis*, eds. Philip Appleman, William A. Madden, and Michael Wolff, 115-30. Bloomington: Indiana University Press, 1959.

Beale, Lionel S. "Pangenesis." Letter to editor. *Nature* 4 (11 May 1871): 25-26.

Beeton, Mary, G.U. Yule, and Karl Pearson. "Data for the Problem of Evolution in Man. V. On the Connection between Duration of Life and Number of Offspring." *Proceedings of the Royal Society of London* 67 (31 October 1900): 159-79.

Beeton, Mary, and Karl Pearson. "Data for the Problem of Evolution in Man. II. A First Study of the Inheritance of Longevity and the Selective Death-rate in Man." *Proceedings of the Royal Society of London* 65 (7 October 1899): 290-305.

Brady, Alexander. "Introduction." In *Essay on Politics and Society*, by John Stuart Mill, ed. J.M. Robson, vol.1, ix-ixx. Toronto: University of Toronto, 1977.

Brooks, W.K. "Francis Galton on the Persistency of Type." *American Journal of Psychology* 1 (November 1887): 173-79.

[Brownson, Orestes A.] "Hereditary Genius." Review of *Hereditary Genius: An Inquiry into Its Laws and Consequences*, by Francis Galton. *Catholic World*, September 1870, 721-32.

Burkhardt, Richard W., Jr. "Closing the Door on Lord Morton's Mare: The Rise and Fall of Telegony." *Studies in the History of Biology* 3 (1979): 1-21.

Burt, Cyril. "Francis Galton and His Contributions to Psychology." *British Journal of Social Psychology* 15 (May 1962): 1-49.

Buss, Allan R. "Galton and the Birth of Differential Psychology and

Eugenics: Social, Political, and Economic Forces." *Journal of the History of the Behavioral Sciences* 13 (January 1976): 47-58.

Campbell, Dudley. "Compulsory Primary Education." *Fortnightly Review*, 1 May 1868, 57-81.

_____. "Compulsory Education." *Westminster Review* n.s. 36 (1 October 1869): 550-56.

Carlyle, Thomas. "Shooting Niagara." *Macmillan's Magazine*, August 1867, 319-36.

"Carranza Lauded in Cooper Union." *New York Times*, 29 June 1916, 4.

Castle, W. E. "The Laws of Heredity of Galton and Mendel, and Some Laws Governing Race Improvement by Selection." *Proceedings of the American Academy of Arts and Sciences* 39 (November 1903): 221-42.

_____. "On the Inheritance of Tricolor Coat in Guinea-Pigs and Its Relation to Galton's Law of Ancestral Heredity." *American Naturalist* 46 (July 1912): 437-40.

Cattell, James McKeen. Review of "The Average Contribution of Each Several Ancestor to the Total Heritage of the Offspring," by Francis Galton. *Psychological Review* 4 (November 1897): 676-77.

_____. "Statistics of American Psychologists." *American Journal of Psychology* 14 (April 1903): 310-28.

_____. "A Statistical Study of American Science. III. The Distribution of American Men of Science." *Science* n.s. 24 (7 December 1906): 732-42.

_____. "Families of American Men of Science." *Popular Science Monthly* 86 (May 1915): 504-15.

Churchill, Frederick B. "From Hereditary Theory to *Vererbung*: The Transmission Problem, 1850-1915." *Isis* 78 (September 1987): 337-64.

Cobbe, Frances Power. "Hereditary Piety." Review of *Hereditary Genius: An Inquiry into Its Laws and Consequences*, by Francis Galton. *Theological Review: A Journal of Religious Thought and Life*, 29 April 1870, 211-34.

_____. "Hereditary Piety." Review of *Hereditary Genius: An Inquiry into Its Laws and Consequences*, by Francis Galton. In *Darwinism in*

Morals, and Other Essays. London: Williams and Norgate, 1872.

Cock, A.G. "William Bateson, Mendelism, and Biometry." *Journal of the History of Biology* 6 (Spring 1973): 1-36.

Cockerell, T.D.A. "The Inheritance of Mental Characters." Letter to editor. *Nature* 65 (January 1902): 245.

_____. "Zoology in America." *Popular Science Monthly* 62 (December 1902): 163-66.

Cole, Henry. "On the International Results of the Exhibition of 1851." In *Lectures on the Results of the Great Exhibition of 1851, Delivered Before the Royal Society of Arts, Manufactures, and Commerce at the Suggestion of H.R.H. Prince Albert, President of the Society*, 2d series, 419-51. London: David Bogne, 1853.

Conklin, Edwin G. "Phenomena of Inheritance." *Popular Science Monthly* 85 (October 1914): 313-37.

Cooley, Charles H. "Genius, Fame and the Comparison of Races." *Annals of the American Academy of Political and Social Science* 9 (May 1897): 317-58.

Cowan, Ruth L. Schwartz. Introduction to *English Men of Science: Their Nature and Nurture*, by Francis Galton. London: Frank Cass, 1970.

_____. "Sir Francis Galton and the Continuity of Germ-Plasm: A Biological Idea with Political Roots." *XIIe Congrès International D'Histoire des Sciences (Paris 1968)*. Paris: Albert Blanchard, 1971, 181-86.

_____. "Francis Galton's Contributions to Genetics." *Journal of the History of Biology* 5 (Fall 1972): 389-412.

_____. "Francis Galton's Statistical Ideas: The Influence of Eugenics." *Isis* 63 (December 1972): 509-28.

_____. "Nature and Nurture: The Interplay of Biology and Politics in the Work of Francis Galton." *Studies in the History of Biology* 1 (1977): 133-208.

Crackanthorpe, Montague. "Sir Francis Galton, F. R. S., A Memoir." *Eugenics Review* 3 (April 1911): 1-9

"Critical Notices: Some Books of the Month." Review of *Hereditary Genius: An Inquiry into Its Laws and Consequences*, by Francis Galton. *Fortnightly Review*, 1 February 1870, 255.

Darden, Lindley. "William Bateson and the Promise of Mendelism." *Journal of the History of Biology* 10 (Spring 1977): 87-106.

Darwin, Charles. "Pangenesis." Letter to editor. *Nature* 4 (27 April 1871): 502.

Darwin, Francis. "Francis Galton, 1822-1911." *Eugenics Review* 6 (April 1914): 1-16.

Davenport, Charles Benedict. "A History of the Development of the Quantitative Study of Variation." *Science* n.s. 12 (7 December 1900): 864-70.

_____. "Light Thrown by the Experimental Study of Heredity upon the Factors and Methods of Evolution." *American Naturalist* 46 (March 1912): 129-38.

Davies, George R. "A Statistical Study of the Influence of the Environment." *Quarterly Journal of the University of North Dakota* 4 (April 1914): 212-36.

de Marrais, Robert. "The Double-Edged Effect of Sir Francis Galton: A Search for Motives in the Biometrician-Mendelian Debate." *Journal of the History of Biology* 7 (Spring 1974): 141-74.

Desmond, Adrian. "Lamarckism and Democracy: Corporations, Corruption and Comparative Anatomy in the 1830s." In *History, Humanity and Evolution: Essays for John C. Greene*, ed. James R. Moore, 99-130. Cambridge: Cambridge University Press, 1989.

D., F.A. "Inheritance of Psychical and Physical Characters in Man." *Nature* 70 (9 June 1904): 137-38.

Diamond, Solomon. "Francis Galton and American Psychology." In *Psychology: Theoretical-Historical Perspectives,* eds. R.W. Rieber and Kurt Salzinger, 43-55. New York: Academic Press, 1980.

Dicey, Edward. "Lincolniana." *Macmillan's Magazine,* June 1865, 185-92.

"Discussion" of "A Practicable Eugenic Solution" by William McDougall. *Sociological Papers* 3 (1907): 81-104.

"Endowed Schools." *Westminster and Foreign Quarterly Review* 21 (1 April 1862): 340-57.

Ewart, J. Cossar. "Opening Address by Prof. J. Cossar Ewart, M.D., F.R.S., President of the Section." *Nature* 64 (12 September 1901): 482-88

Fancher, Raymond E. "Alphonse de Candolle, Francis Galton, and the

Early History of the Nature-Nurture Controversy." *Journal of the History of the Behavioral Sciences* 19 (October 1983): 341-52.

_____. "Francis Galton's African Ethnography and Its Role in the Development of His Psychology." *British Journal for the History of Science* 16 (1983): 67-79.

Farrall, Lyndsay A. "Controversy and Conflict in Science: A Case Study—The English Biometric School and Mendel's Laws." *Social Studies of Science* 5 (1975): 269-301.

Farrar, Rev. F.W. "Hereditary Genius." Review of *Hereditary Genius: An Inquiry into Its Laws and Consequences*, by Francis Galton. *Fraser's Magazine*, August 1870, 251-65.

Fawcett, Cicely D. and Karl Pearson. "Mathematical Contributions to the Theory of Evolution. On the Inheritance of the Cephalic Index." *Proceedings of the Royal Society of London* 62 (16 March 1898): 413-17.

Froggatt, P. and N.C. Nevin. "Galton's 'Law of Ancestral Heredity': Its Influence on the Early Development of Human Genetics." *History of Science* 10 (1971): 1-27.

_____. "The 'Law of Ancestral Heredity' and the Mendelian-Ancestrian Controversy in England, 1889-1906." *Journal of Medical Genetics* 8 (March 1971): 1-36.

Galton, Francis. "Recent Expedition into the Interior of South-Western Africa." *Journal of the Royal Geographical Society* 22 (1852): 140-63.

_____. "The First Steps Towards the Domestication of Animals." *Transactions of the Ethnological Society of London* 3 (1865): 122-38.

_____. "Hereditary Talent and Character." *Macmillan's Magazine*, "Part I," June 1865, 157-66; "Second Paper," August 1865, 318-27.

_____. "Hereditary Genius: The Judges of England between 1660 and 1865." *Macmillan's Magazine*, March 1869, 424-31.

_____. "Pangenesis." Letter to editor. *Nature* 4 (4 May 1871): 5-6.

_____. "Experiments in Pangenesis, by Breeding from Rabbits of a Pure Variety, into Whose Circulation Blood Taken from Other Varieties Had Previously Been Largely Transfused." *Proceedings of the Royal Society of London* 19 (16 June 1870 to 15 June 1871): 393-410.

_____. "On Blood Relationship." *Proceedings of the Royal Society of London* 20 (13 June 1872): 394-402.

_____. "Hereditary Improvement." *Fraser's Magazine*, January 1873, 116-30.

_____. "On the Causes Which Operate to Create Scientific Men." *Fortnightly Review*, 1 March 1873, 345-51.

_____. "On Men of Science, Their Nature and Nurture." *Proceedings of the Royal Institution* 7 (1874): 227-36.

_____. "The History of Twins as a Criterion of the Relative Powers of Nature and Nurture." *Fraser's Magazine*, November 1875, 566-76.

_____. "A Theory of Heredity." *Contemporary Review* 27 (December 1875): 80-95.

_____. "The History of Twins as a Criterion of the Relative Powers of Nature and Nurture." *Journal of the Anthropological Institute of Great Britain and Ireland* 5 (January 1876): 391-406.

_____. "Short Notes on Heredity &c in Twins." *Journal of the Anthropological Institute of Great Britain and Ireland* 5 (January 1876): 324-29.

_____. "A Theory of Heredity." *Journal of the Anthropological Institute of Great Britain and Ireland* 5 (January 1876): 329-48.

_____. "Typical Laws of Heredity." *Notices of the Proceedings of the Meetings of the Members of the Royal Institution of Great Britain with Abstract of the Discourses Delivered at the Evening Meetings* 8 (1875-1878): 282-301.

_____. "Types and Their Inheritance." *Science* 6 (25 September 1885): 268-74.

_____. "Heredity in Business Merit." *Journal of Heredity* 1 (October 1885): 4-6.

_____. "Regression Towards Mediocrity in Hereditary Stature." *Journal of the Anthropological Institute of Great Britain and Ireland* 15 (November 1885): 246-63.

_____. President's Address. *Journal of the Anthropological Institute of Great Britain and Ireland* 15 (1886): 489-99.

_____. "Family Likeness in Stature." *Proceedings of the Royal Society of London* 40 (21 January 1886): 42-63.

_____. "Hereditary Stature." Letter to editor. *Nature* 33 (4 February 1886): 317.

_____. "Family Likeness in Eye-Colour." *Proceedings of the Royal Society of London* 40 (27 May 1886): 402-16.

_____. "The Patterns in Thumb and Finger Marks.—On Their Arrangement into Naturally Distinct Classes, the Permanence of the Papillary Ridges that Make Them, and the Resemblance of Their Classes to Ordinary Genera." *Philosophical Transactions of the Royal Society of London* 182B (1891): 1-23.

_____. "Discontinuity in Evolution." *Mind* n.s. 19 (July 1894): 362-72.

_____. "Rate of Racial Change that Accompanies Different Degrees of Severity in Selection." Letter to editor. *Nature* 55 (29 April 1897): 605-06.

_____. "The Average Contribution of Each Several Ancestor to the Total Heritage of the Offspring." *Proceedings of the Royal Society of London* 61 (31 July 1897): 401-13.

_____. "Hereditary Colour in Horses." *Nature* 56 (21 October 1897): 598-99.

_____. "The Distribution of Prepotency." Letter to editor. *Nature* 58 (14 July 1898): 246-47.

_____. "The Possible Improvement of the Human Breed Under the Existing Conditions of Law and Sentiment." *Annual Report of the Board of Regents of the Smithsonian Institution* (1901): 523-38.

_____. "The Possible Improvement of the Human Breed Under the Existing Conditions of Law and Sentiment." *Nature* 64 (31 October 1901): 659-65.

_____. "On the Probability that the Son of a Very Highly-Gifted Father Will Be No Less Gifted." Letter to editor. *Nature* 65 (28 November 1901): 79.

_____. "The Possible Improvement of the Human Breed Under the Existing Conditions of Law and Sentiment." *Popular Science Monthly* 60 (January 1902): 218-33.

_____. "Eugenics." Letter to editor. *Westminster Gazette*, 26 June 1908, 1-2.

_____. "Alcoholism and Offspring." Letter to editor. *Times* [of London], 3 June 1910, 6.

_____. "The Eugenics Laboratory and the Eugenics Education Society."

Letter to editor. *Times* [of London], 2 November 1910, 9.

"Galton's Development of the Human Faculty." Review of *Inquiries into the Human Faculty*, by Francis Galton. *Nation*, 14 June 1883, 512-13.

Geison, Gerald L. "Darwin and Heredity: The Evolution of His Hypothesis of Pangenesis." *Journal of the History of Medicine and Allied Sciences* 24 (October 1969): 375-411.

Giddings, Franklin H. "Darwinism in the Theory of Social Evolution." *Popular Science Montly* 75 (July 1909): 75-89.

Gökyigit, Emel Aileen. "The Reception of Francis Galton's *Hereditary Genius* in the Victorian Periodical Press." *Journal of the History of Biology* 27 (1994): 215-40.

Greene, John C. "Darwin as a Social Evolutionist." *Journal of the History of Biology* 10 (Spring 1977): 1-27.

Haines, George IV. "Technology and Liberal Education." In *1859: Entering an Age of Crisis*, eds. Philip Appleman, William A. Madden, and Michael Wolff, 97-112. Bloomington: Indiana University Press, 1959.

Halliday, R.J. *John Stuart Mill*. London: Allen and Unwin, 1976.

Hargitt, Charles W. "A Problem in Educational Eugenics." *Popular Science Monthly* 83 (October 1913): 355-67.

Harris, George. "Hereditary Genius." Review of *Hereditary Genius: An Inquiry into Its Laws and Consequences*, by Francis Galton. *Journal of Anthropology* 1 (July 1870): 56-65.

Harrison, Frederic. "The Revival of Authority." *Fortnightly Review*, 1 January 1873, 1-26.

"Hereditary Genius." Review of *Hereditary Genius: An Inquiry into Its Laws and Consequences*, by Francis Galton. *Appleton's Journal of Popular Literature, Science, and Art*, 19 February 1870, 217-18.

"Hereditary Genius." Review of *Hereditary Genius: An Inquiry into Its Laws and Consequences*, by Francis Galton. *Atlantic Monthly*, June 1870, 753-56.

"Hereditary Genius." Review of *Hereditary Genius: An Inquiry into Its Laws and Consequences*, by Francis Galton. *Saturday Review*, 25 December 1869, 832-33.

Hilts, Victor L. "William Farr (1807-1883) and the Human Unit." *Victorian*

Studies 14 (December 1970): 143-50.

"History and Biography." Review of *Hereditary Genius: An Inquiry into Its Laws and Consequences*, by Francis Galton. *Westminster and Foreign Quarterly Review* 93, (1 January 1870), 296-313.

Hodson, Kathleen. "The Eugenics Review 1909-1968." *Eugenics Review* 60 (September 1968): 162-75.

Holley, Charles Elmer. "The Relationship between Persistence in School and Home Conditions." In *Fifteenth Yearbook of the National Society for the Study of Education* (1916), Part II, 8-119. Chicago University Press, 1916.

Huxley, Julian. "The Case for Eugenics." *Sociological Review* 18 (October 1926): 279-90.

Huxley, Thomas. "A Liberal Education; And Where to Find It. An Inaugural Address." *Macmillan's Magazine*, March 1868, 367-78.

Jennings, H.S. "Experimental Evidence of the Effectiveness of Selection." *American Naturalist* 44 (March 1910): 136-45.

Judd, John W. Letter to editor. *Nature* 85 (26 January 1911): 405-06.

_____. Letter to editor. *Nature* 85 (9 February 1911): 474-75.

Keith, Arthur. "Galton's Place Among Anthropologists." *Eugenics Review* 12 (April 1920), 14-28.

Kellogg, John H. "Tendencies Toward Race Degeneracy." *New York Medical Journal* 94 (2 September 1911): 461-67; (9 September 1911): 526-29.

Kern, Paul B. "Universal Suffrage without Democracy: Thomas Hare and John Stuart Mill." *Review of Politics* 34 (July 1972): 306-22.

Kevles, Daniel J. "Annals of Eugenics: A Secular Faith — I." *New Yorker*, 8 October 1984, 51-52, 55, 56, 59-60, 62, 64-68, 70, 75, 76, 78-86, 89-90, 93-96, 98-110, 113-15.

_____. "Genetics in the United States and Great Britain 1890-1930: A Review with Speculations." In *Biology, Medicine and Society, 1840-1940*, ed. Charles Webster, 193-215. Cambridge: Cambridge University Press, 1981.

Kolbe, Rev. F.N. "An Account of the Damara Country." *Journal of the Ethnographic Society of London* 3 (1854): 1-3.

Kropotkin, P. "Inheritance of Acquired Characters: Theoretical Difficulties." *Nineteenth Century and After*, March 1912, 511-31.

[Lankester, E. Ray]. "Darwin Versus Lamarck." *Nature* 39 (28 February 1889): 428-29.

_____. Letter to editor, *Nature* 41 (6 March 1890): 415-16.

_____. Letter to editor, *Nature* 41 (27 March 1890): 486-88.

Larrabee, W.H. "De Candolle on the Production of Men of Science." *Popular Science Monthly* 29 (May 1886): 34-46.

Lock, R. H. "Methods of Research, Being an Attempt to Reconcile the Views of Dr. G. Archdall Reid with Those of Other Biologists." *Eugenics Review* 3 (January 1911): 340-45.

MacKenzie, Donald. "Eugenics in Britain." *Social Studies of Science* 6 (September 1976): 499-532.

_____. "Statistical Theory and Social Interests: A Case Study." *Social Studies of Science* 8 (February 1978): 35-83.

_____. "Karl Pearson and the Professional Middle Class." *Annals of Science* 36 (March 1979): 125-43.

_____. "Sociobiologies in Competition: The Biometrician-Mendelian Debate." In *Biology, Medicine and Society 1840-1940*, ed. Charles Webster, 243-88. Cambridge: Cambridge University Press, 1981.

MacKenzie, D., and S.B. Barnes. "Biometriker versus Mendelianer. Eine Kontroverse und ihre Erklarung." *Kölner Zeitschrift für Soziologie und Sozialpsychologie*. Sonderheft 18 (1975), 165-96.

MacKenzie, Donald, and Barry Barnes. "Scientific Judgement: The Biometry-Mendelism Controversy." In *Natural Order: Historical Studies of Scientific Culture*, eds. Barry Barnes and Steven Shapin, 191-210. Beverly Hills, CA: Sage Publications, 1979.

McDougall, William. "A Practicable Eugenic Solution." *Sociological Papers* 3 (1907): 55-80.

_____. Letter to editor. *New Republic*, 27 June 1923, 125-26.

Medola, R. "A Contribution to the History of Evolution." *Nature* 85 (5 January 1911): 297-98.

[Merivale, Herman.] "Hereditary Genius: An Inquiry into Its Laws and Consequences." Review of *Hereditary Genius: An Inquiry into Its*

Laws and Consequences, by Francis Galton. *Edinburgh Review*, July 1870, 100-25.

"Miss Meteyard's Life of Wedgwood." Review of *Life of Josiah Wedgwood, from his Private Correspondence and Family Papers; With an Introductory Sketch of the Art of Pottery in England*, by E. Meteyard. *Macmillan's Magazine*, June 1865, 155-57.

Mill, John Stuart. "Recent Writers on Reform." In Mill, John Stuart. *Essays on Politics and Society*, ed. J.M. Robson. Vol. 2, 341-70. Toronto: University of Toronto Press, 1977.

Morgan, T. H. "For Darwin." *Popular Science Monthly* 74 (April 1909): 367-80.

_____. "Human Inheritance." *American Naturalist* 58 (September-October 1924): 385-409.

Morley, John. "Social Responsibilities." *Macmillan's Magazine*, September 1866, 377-86.

"A New Law of Heredity." *Nature* 56 (8 July 1897): 235-37.

Norton, Bernard J. "The Biometric Defense of Darwinism." *Journal of the History of Biology* 6 (Fall 1973): 283-316.

_____. "Biology and Philosophy: The Methodological Foundations of Biometry." *Journal of the History of Biology* 8 (Spring 1975): 85-93.

_____. "Metaphysics and Population Genetics: Karl Pearson and the Background to Fisher's Multi-factorial Theory of Inheritance." *Annals of Science* 32 (1975): 537-53.

_____. "Karl Pearson and Statistics: The Social Origins of Scientific Innovation." *Social Studies of Science* 8 (February 1978): 3-34.

_____. "Psychologists and Class." In *Biology, Medicine and Society 1840-1940*, ed. Charles Webster, 289-314. Cambridge: Cambridge University Press, 1981.

"Of Success in Life." *Macmillan's Magazine*, June 1865, 181-85.

Olby, Robert. "The Dimensions of Scientific Controversy: The Biometric-Mendelian Debate." *British Journal for the History of Science* 22 (September 1989): 299-320.

Owen, G.E.L. "Dialectic and Eristic in the Treatment of Forms." In *Aristotle on Dialectic--The Topics*, ed. G.E.L. Owen, 103-07. Oxford: Oxford University Press, 1968.

"The Paris Exhibition and Industrial Education." *Journal of the Society of Arts*, 6 July 1867, 477-79.

Parkyn, E.A. Letter to editor. *Nature* 85 (9 February 1911): 474.

Pearson, Egon S. "Karl Pearson: An Appreciation of Some Aspects of His Life and Work. Part I: 1857-1906." *Biometrika* 28 (December 1936): 193-257.

_____. "Karl Pearson: An Appreciation of Some Aspects of His Life and Work. Part II: 1906-1936." *Biometrika* 29 (February 1938): 161-248.

Pearson, Karl. "Anarchy." *Cambridge Review*, 30 March 1881, 268-70.

_____. "Contributions to the Mathematical Theory of Evolution. – II. Skew Variations in Homogeneous Material." *Philosophical Transactions of the Royal Society of London* 186A (1895): 343-414.

_____. "Contributions to the Mathematical Theory of Evolution. Note on Reproductive Selection." *Proceedings of the Royal Society of London* 59 (18 June 1896): 301-05.

_____. "Mathematical Contributions to the Theory of Evolution. – III. Regression, Heredity and Panmixia." *Philosophical Transactions of the Royal Society of London* 187A (1896): 253-318.

_____. "Mathematical Contributions to the Theory of Evolution. On the Law of Ancestral Heredity." *Nature* 57 (10 March 1898): 452-53.

_____. "On the Law of Ancestral Heredity." *Science*, n.s. 7 (11 March 1898): 337-39.

_____. "Mathematical Contribution to the Theory of Evolution. On the Law of Ancestral Heredity." *Proceedings of the Royal Society of London* 62 (16 March 1898): 386-412.

_____. "Mathematical Contributions to the Theory of Evolution. On the Law of Reversion." *Proceedings of the Royal Society of London* 66 (22 March 1900): 140-64.

_____. "Data for the Problem of Evolution in Man. IV. Note on the Effect of Fertility Depending on Homogamy." *Proceedings of the Royal Society of London* 66 (12 May 1900): 316-23.

_____. "On the Inheritance of Mental Characters in Man." *Proceedings of the Royal Society of London* 69 (24 December 1901): 153-55.

_____. "Mathematical Contributions to the Theory of Evolution. – VII. On the Inheritance of Characters Not Quantitatively Measurable."

Philosophical Transactions of the Royal Society of London 195A (1901): 1-47.

_____. "The Inheritance of Mental Characters." Letter to editor. *Nature* 65 (16 January 1902): 245-46.

_____. "The Inheritance of Mental Characters." Letter to editor. *Nature* 65 (27 February 1902): 391.

_____. "Prefatory Essay: The Function of Science in the Modern State." *Encyclopedia Britannica.* 10th ed. Vol. 32; vol. 8 of new vols. Edinburgh: Adam and Charles Black, 1902, vii-xxxvii.

_____. "On the Inheritance of the Mental and Moral Characters in Man, and Its Comparison with the Inheritance of the Physical Characters." *Journal of the Anthropological Institute of Great Britain and Ireland* 33 (July-December 1903): 179-237.

_____. "On a Criterion Which May Serve to Test Various Theories of Inheritance." *Proceedings of the Royal Society of London* 73 (7 May 1904): 262-80.

[_____.] "Francis Galton. February 16, 1822-January 17, 1911." *Nature* 85 (2 February 1911): 440-45.

Pearson, Karl, and Alice Lee. "Mathematical Contributions to the Theory of Evolution. On Telegony in Man, &c." *Proceedings of the Royal Society of London* 60 (9 December 1896): 273-83.

_____. "Mathematical Contributions to the Theory of Evolution. On the Relative Variation and Correlation in Civilised and Uncivilised Races." *Proceedings of the Royal Society of London* 61 (5 July 1897): 343-57.

_____. "Mathematical Contributions to the Theory of Evolution. – VIII. On the Inheritance of Characters Not Capable of Exact Quantitative Measurement. Part I, Introductory; Part II, On the Inheritance of Coat Colour in Horses; Part III, On the Inheritance of Eye Colour in Man." *Philosophical Transactions of the Royal Society of London* 195A (1901): 79-150.

_____. "On the Laws of Inheritance in Man." *Biometrika* 2 (November 1903): 357-462.

Pearson, Karl, Alice Lee, Ernest Warren, Agnes Fry, Cicely D. Fawcett and Others. "Mathematical Contributions to the Theory of Evolution. – IX. On the Principle of Homotyposis and Its Relation to Heredity, to the Variability of the Individual and to That of the Race Part I. "Homotyposis in the Vegetable Kingdom." *Philosophical Transactions of the Royal Society of London* 197A (1901): 285-379.

Pearson, Karl, Alice Lee, and Leslie Bramley-Moore. "Mathematical Contributions to the Theory of Evolution. — VI. Genetic (Reproductive) Selection: Inheritance of Fertility in Man, and of Fecundity in Throughbred Race-horses." *Philosophical Transactions of the Royal Society of London* 192A (1899): 257-330.

Pearson, Karl, and L.N.G. Filon. "Mathematical Contributions to the Theory of Evolution. — IV. On the Probable Errors of Frequency Constants and on the Influence of Random Selection on Variation and Correlation." *Philosophical Transactions of the Royal Society of London* 191A (1898): 229-311.

Pearson, Karl (with appendices by various authors). "The Law of Ancestral Heredity." *Biometrika* 2 (February 1903): 211-34.

"Popular Education." *Westminster and Foreign Quarterly Review* n.s. 33 (1 April 1868): 421-41.

"Popular Education in Prussia." *Westminster and Foreign Quarterly Review* 21 (1 January 1862): 169-200.

"Popular Education: The Means of Obtaining It." *Prospective Review: A Quarterly Journal of Theology and Literature* 8 (1852): 16-56.

"Primary Education." *Westminster Review* n.s. 35 (1 April 1869): 458-83.

"Questions Bearing on Specific Ability." *Nature* 51 (11 April 1895): 570-71.

Reid, G. Archdall. "Methods of Research." *Eugenics Review* 3 (October 1911): 241-64.

_____. "The Inheritance of Mental Characters." Letter to editor. *Nature* 88 (14 December 1911): 210-11.

"Report of the Commissioners Appointed to Inquire into the State of Popular Education in England. Vol. I. Presented to Parliament in April, 1861." *British Quarterly Review* 34 (July 1861): 218-33.

"Report of the Commissioners Appointed to Inquire into the State of Popular Education in England. London 1861." *London Review*, July 1861, 503-35.

Review of *Hereditary Genius: Its Laws and Consequences*, by Francis Galton.

Galaxy, March 1870, 424.

Review of *Hereditary Genius: Its Laws and Consequences*, by Francis Galton. *Harper's New Monthly Magazine*, May 1870, 928-29.

Robertson, John M. "The Economics of Genius." *Forum* 25 (April 1898): 178-90.

Roll-Hansen, Nils. "The Controversy between Biometricians and Mendelians: A Test Case for the Sociology of Knowledge." *Social Science Information* 19 (1980): 501-17.

Russell, J. Scott. "Technical Education a National Want." *Macmillan's Magazine*, April 1868, 447-59.

Sapp, Jan. "The Struggle for Authority in the Field of Heredity, 1900-1932: New Perspectives on the Rise of Genetics." *Journal of the History of Biology* 16 (Fall 1983): 311-34.

Schenk, Faith, and A.S. Parkes. "The Activities of the Eugenics Society." *Eugenics Review* 60 (September 1968): 142-61.

Searles, G.R. "Eugenics and Class." In *Biology, Medicine and Society*, ed. Charles Webster, 217-42. Cambridge: Cambridge University Press, 1981.

Semmel, Bernard. "Karl Pearson: Socialist and Darwinist." *British Journal of Sociology* 9 (June 1958): 111-25.

Slater, Eliot. "Galton's Heritage." *Eugenics Review* 52 (July 1960): 91-103.

Stanley, Hiram M. "Mr. Galton on Natural Inheritance." Letter to editor. *Nature* 40 (31 October 1889): 642-43.

Stephens, Michael D. and Gordon W. Roderick. "British Artisan Scientific and Technical Education in the Early Nineteenth Century." *Annals of Science* 29 (June 1972): 87-98.

_____. "Science, the Working Classes and Mechanics' Institutes." *Annals of Science* 29 (December 1972): 349-60.

_____. "Changing Attitudes to Education in England & Wales 1833-1902: The Governmental Reports, with Particular Reference to Science & Technical Studies." *Annals of Science* 30 (June 1973): 149-64.

_____. "American and English Attitudes to Scientific Education During the Nineteenth Century." *Annals of Science* 30 (December 1973): 435-56.

Suvin, Darko. "The Social Addressees of Victorian Fiction: A Preliminary Inquiry." *Literature and History* 8 (Spring 1982): 11-40.

Swinburne, R.G. "Galton's Law—Formulation and Development." *Annals of Science* 21 (March 1965): 15-31.

Terman, Lewis M. "Genius and Stupidity: A Study of Some of the Intellectual Processes of Seven 'Bright' and Seven 'Stupid' Boys." *Pedagogical Seminary* 13 (September 1906): 307-73.

_____. "The Significance of Intelligence Tests for Mental Hygiene." *Journal of Psycho-Asthenics* 18 (March 1914): 119-27.

_____. "The Significance of Intelligence Tests for Mental Hygiene." *Transactions, International Congress on School Hygiene* 3 (1914): 502-08.

_____. "The Psychological Determinist; Or Democracy and the I.Q." *Journal of Educational Research* 6 (June 1922): 57-62.

_____. "Were We Born That Way? Or Can We Help It? Is Heredity of Environment the Power that Moulds Us? What Science Now Knows About Intellectual Differences, and Their Significance." *World's Work*, October 1922, 655-60.

_____. "Lewis M. Terman [,] Trails to Psychology." In *A History of Psychology in Autobiography*, vol. 2, ed. Carl Murchison, 297-331. Worcester, MA: Clark University Press, 1932.

Thiselton-Dyer, W.T. Letter to editor. *Nature* 85 (26 January 1911): 405-06.

Thomson, J. Arthur. "Sir Francis Galton." *Sociological Review* 4 (April 1911): 141-42.

Thorndike, Edward L. "Marriage Among Eminent Men." *Popular Science Monthly* 61 (August 1902): 328-29.

_____. "Edward Lee Thorndike." In *A History of Psychology in Autobiography*, vol. 3, ed. Carl Murchison, 263-70. Worcester, MA: Clark University Press, 1936.

Turner, William. "Opening Address by Prof. Sir William Turner, M.B., LL.D., FR.SS.L.&E., President of the Section." *Nature* 40 (26 September 1889): 526-33.

Vorzimmer, Peter J. "Darwin's 'Lamarckism' and the 'Flat-Fish Controversy' (1863-1871)." *Lychnos* 1969-1970: 121-70.

Wallace, Alfred Russel. "Hereditary Genius." Review of *Hereditary Genius:*

An Inquiry into Its Laws and Consequences, by Francis Galton. *Nature* 1, 17 March 1870, 501-03.

Warren, Charles. "The Failure of the Democratic Idea in City Government." Commencement Dissertation, June 1889. Harvard University Archives. Cambridge, Massachusetts.

_____. "Plato's Republic with Reference Especially to the Functions of the State Therein and Its Relation with Some Modern Questions." Essay in Philosophy I, 1888-89, 9 February 1889. Harvard University Archives. Cambridge, Massachusetts.

Wilkie, J.S. "Galton's Contribution to the Theory of Evolution with Special Reference to His Use of Models and Metaphors." *Annals of Science* 11 (September 1956): 194-205.

Yerkes, Robert M. "Eugenics: Its Scientific Basis and Its Program." Address Before the Eugenics Section of the National Conference of Charities and Correction, given on 19 June 1912, Cleveland, Ohio. Abstracted in *Proceedings of the National Conference of Charities and Correction* 39 (1912): 279-80.

_____. "Eugenics: Its Scientific Basis and Its Program." Address Before the Eugenics Section of the National Conference of Charities and Correction, given on 19 June 1912. Robert M. Yerkes Papers, Manuscripts and Archives, Sterling Memorial Library, Yale University. New Haven, Connecticut.

INDEX

Pearson, Karl
works of, cited, continued:
Inheritance in Man" (1903), 74
n.33, 76
"On a Criterion Which May Serve
to Test Theories of Inheritance"
(1904), 74 n.33
"Francis Galton. February 16,
1822- January 17, 1911" (1911),
19 n.2, 29 n.16, 75 n.33
*The Life, Letters and Labours of
Francis Galton*, vol.1 (1914), 19-
20, 20 n.3, 21 n.5, 22 n.8, 23 nn.9,
10, 24 n.11, 25 n.12, 26 nn.13, 14,
47 n.25, 57 n.5
*Francis Galton 1822-1922: A
Centenary Appreciation* (1922), 20
n.2, 56 n.5, 57 n.5
*The Life, Letters and Labours of
Francis Galton*, vol. 2 (1924), 8
n.7, 15 n.17, 16 n.18, 17 n.19, 18,
19, 21 n.5, 26 n.14, 28 n.15, 47 n.25,
55 n.4, 57 n.5, 59 n.8, 60 n.11, 63
n.20, 64 n.21, 64 nn.22, 23, 24, 25, 66
n.27, 77 n.36, 84 n.44
*The Life, Letters and Labours of
Francis Galton*, vol. 3A (1930), 55
n.4, 66 n.27, 87 n.49, 89 nn.49,
50
*The Life, Letters and Labours of
Francis Galton*, vol. 3B (1930), 17
n.19, 28 n.15, 55 n.4, 87 n.49
*The History of Statistics in the 17th
and 18th Centuries Against the
Changing Background of
Intellectual, Scientific and
Religious Thought: Lectures by
Karl Pearson Given at University
College London during the
Academic Sessions 1921-1933*
(1978), 32 n.2
Perkin, Harold, 41
Plato, 33-35
American antidemocratic admirers
of, 46-47
English opponents of democracy
linked with, 46
Galton and, Yerkes on, 33
Playfair, Lyon, 36 n.8, 38 n.9
political context of Galton's invention of
eugenics, 32-49
Political Progress Not Necessarily Democratic
(Lorimer), 43-44

political purpose in eugenics, 32-35, 101-
102 n.6, 104
perceptible indications of, 11, 43 n.
13, 47-49, 50-92
Poole's Index to Periodical Literature, 97
Popenoe, Paul, 18 n.22
Porson, Richard, 10
Positivists, French and English, 46
"Practicable Eugenic Suggestion"
(McDougall), 2
"prenatal culture," 53
Priestley, Joseph, 20
Proceedings of the Royal Society, 64
psychohistorical explanations of Galton's
invention of eugenics, 28-31

regression to the mean, 51, 68
Galton's theoretical statements on, 72
n.32
as inducement for oligarchy, 54-55
jettisoned by Pearson, 71
Pearson's further views on, 71 n.32, 74-
75 n.33
readmitted by Pearson, 73, 74-75 n.33
Reader, The, 59
co-owned by Galton, 28
description of, 28 n.15
Republic (Plato), 33-35
Royal Geographical Society, 25, 28
Royal Society of London, 25-26, 59, 64
Ruskin, John, 46
Russell, John Scott, 36 n.8, 38 n.9

Sarton, George, 32
Scaliger, Joseph Justus, 9
Scaliger, Julius Caesar, 9
Schuster, Edgar, 86
sequence of introduction of Galton's
tenets, 57
Pearson's claims regarding, 56-57 n.5
Sévigné, de, Marquise, 10
sibling study, Pearson's
criticism of, 80, 80-81 n.39
description of, 78-80, 79 n.38 –81 n.39
Thorndike's contrastant responses to,
99-100
Slaughter, J. W., 86 n.46, 87 n.48
biographical detail on, 87-88 n.49
chairman of Eugenics Education
Society, 87
convicted of indecent assault, 87
conviction overturned, 87

www.ingramcontent.com/pod-product-compliance
Lightning Source LLC
Chambersburg PA
CBHW080928100426
42812CB00007B/2403